ENVIRONMENTAL RESEARCH ADVANCES

MULTIPLE-USE MANAGEMENT OF PROTECTED NATURAL AREAS

INTEGRATING CONSERVATION, RESTORATION AND RECREATION IN THE SPANISH BASQUE COUNTRY

Environmental Research Advances

Additional books in this series can be found on Nova's website under the Series tab.

Additional E-books in this series can be found on Nova's website under the E-book tab.

Environmental Remediation Technologies, Regulations and Safety

Additional books in this series can be found on Nova's website under the Series tab.

Additional E-books in this series can be found on Nova's website under the E-book tab.

ENVIRONMENTAL RESEARCH ADVANCES

MULTIPLE-USE MANAGEMENT OF PROTECTED NATURAL AREAS

INTEGRATING CONSERVATION, RESTORATION AND RECREATION IN THE SPANISH BASQUE COUNTRY

ASHLEY DUVAL,
TAMARA MURUETAGOIENA
AND
MARK S. ASHTON
EDITORS

nova publishers

Nova Science Publishers, Inc.
New York

For permission to use material from this book please contact us:
Telephone 631-231-7269; Fax 631-231-8175
Web Site: http://www.novapublishers.com

LIBRARY OF CONGRESS CATALOGING-IN-PUBLICATION DATA

Multiple-use management of protected natural areas : integrating conservation, restoration, and recreation in the Spanish Basque country / editors, Ashley DuVal, Mark S. Ashton, Tamara Muruetagoiena.
p. cm.
Includes index.
ISBN 978-1-61942-328-2 (hardcover)
1. Natural areas--Spain--Pams Vasco. 2. Restoration ecology--Spain--Pams Vasco. 3. Recreation areas--Spain--Pams Vasco. 4. Natural history--Spain--Pams Vasco. 5. Pams Vasco (Spain) I. DuVal, Ashley. II. Ashton, Mark S. III. Muruetagoiena, Tamara.
QH77.S628M85 2011
333.720946'6--dc23

2011047273

Published by Nova Science Publishers, Inc. † New York

CONTENTS

ABBREVIATIONS

Acer campestre – hedge maple
Alnus glutinosa – black alder
Arbutus unedo – strawberry madrone
Betula celtiberica – Iberian white birch
Betula pendula – silver birch
Castanea sativa – Spanish chestnut
Charaxes jasius –two-tailed pasha
Corylus avellana – common hazel
Crataegus monogyna – common hawthorne
Fagus sylvatica – European beech
Fraxinus excelsior – European ash
Galemys pyrenaicus – Pyrenean Desman
Laurus nobilis – bay laurel
Liriodendron tulipifera – tulip poplar
Lucanus cervus – stag beetle
Malus sylvestris – European wild apple
Osmoderma eremita – hermit beetle
Pinus radiata – Monterey pine
Prunus spinosa - blackthorn
Pseudotsuga menziesii – Douglas-fir
Quercus ilex – holm oak
Quercus pyrenaica – Pyrenean oak
Quercus robur – pedunculate oak
Quercus rubra – red oak
Quercus suber – cork oak
Rhamnus frangula.- alder buckthorn
Rosalia alpina – rosalia longicorn

Salix atrocinerea – large gray willow
Soldanella villosa – hairy snowbell
Sorbus aucuparia – European mountain ash
Ulmus glabra – Wych elm
Ulmus minor – field elm.

ACKNOWLEDGMENTS

This book would not have been possible without the time, efforts, energy and support of many individuals and organizations. In particular, many thanks to the School of Forestry and Environmental Studies for making this course and field visit possible; Professor Mark S. Ashton and Tamara Muruetagoiena for the incredible amount of effort they put towards planning and coordinating our visit to the Basque Country, instructing the course and advising the students; to our gracious host Iñaki Aizpuru for putting together such a fantastic and thorough schedule and giving us all such an excellent taste of Basque culture and hospitality; to the course teaching assistant Javier Fernandez-Vega for his excellent work coordinating details and logistics throughout the phases of the trip and compiling such excellent resources to enhance our visit; to our reviewers Iñaki Aizpuru and Arturo Elosegi for their thoughtful comments and useful insights throughout the development of these papers and to Michele Abbene for photo documentation of the Rapid Assessment that appears throughout this volume. Finally, thanks to the many helpful and knowledgement experts, technicians, rangers, guides, professors and government associates who accompanied us throughout our trip and greatly enhancing our experience and insights. These include Asun Yarzabal, Eneko Garmendia, Jon Ugarte, Kiko Alvarez, Iñaki Azkarate, Ibai Aizpuru, Imanol Arruti, Carlos Aseginolaza, Jose Antonio Belzunce, Alberto Luengo, Jorge Gonzalez, Mikel Zabala, Valentin Mugarza Martinez, Ibai Portu, John Fernandez, Xarles Cepedo, Jabi Zabala Albizua, Giuseppe Traverso and Ibon Tamayo.

In: Multiple-Use Management ... ISBN 978-1-61942-328-2
Eds: A. DuVal, T. Muruetagoiena et al.

INTRODUCTION

***Ashley DuVal*[1], *Tamara Muruetagoiena*[2] *and Mark S. Ashton*[3]**
[1]Yale School of Forestry and Environmental Studies
[2]Yale School of Forestry and Environmental Studies, Southern Europe Forest Owners Association in Spain, France, Italy, and Portugal
[3]Silviculture and Forest Ecology at the Yale School of Forestry and Environmental Studies

ABSTRACT

In the fall of 2009, students at the Yale School of Forestry and Environmental Studies along with members of the Euskadi Biodiversity Center (Spain) participated in a field appraisal to better understand and respond to conservation and management concerns in Aiako Harria Natural Park (Basque Country, Spain). The analyses and recommendations in this volume are the product of a semester-long case study course intended to promote professional development through interdisciplinary problem solving utilizing the rapid assessment technique. Drawing upon prior experience in conservation and management, case studies in the literature and first hand observations made in our field visit to Aiako Harria, this series of papers outlines actions for the Euskadi Biodiversity Center and other Protected Natural Area managers to consider in addressing site specific land management challenges, particularly in Europe. Each chapter represents a different but complementary focus of land management within the park. A multidisciplinary approach is taken to protected area management, drawing from disciplinary perspectives in policy science, environmental science, political ecology and silviculture to address the historical, socio-political, economic and biophysical origins of contemporary conflicts in resource use and management.. This introduction provides a sociocultural and biophysical background of the region, as well as a description of the sites visited and activities undertaken during the rapid assessment. A brief summary of the topics and main points discussed in each paper follows.

Keywords: Aiako Harria, natural resource management, biodiversity, autonomous Basque region, rapid assessment, ecosystem services

INTRODUCTION

This monograph uses the case study of Aiako Harria Natural Park in the Basque region of Northern Spain to address key issues and opportunities in the integrative management of public lands. The articles that follow are based on the outcomes of a three-stage rapid assessment project conducted over the course of several months in the fall of 2009, including a week-long intensive field visit. The Rapid Assessments in Forest Conservation course is intended to develop professionalism in graduate students through practical, applied skill development in multiple phases of problem solving (Clark and Ashton 2004). The course has been part of the Yale School of Forestry and Environmental Studies curriculum since 1993 and has in past years included rapid appraisals in Ecuador, Belize, Mexico, Venezuela, Panama, Costa Rica and Hawaii. This was the first field course to conduct a case study on a temperate European forest. In the introduction we provide a description of the biophysical, natural, socio-political and historical characteristics of the Basque country, outline and discuss the conservation priorities articulated by park managers and summarize the students papers addressing these resource management priorities.

Regional Background

Basque Country refers to a territory in the Atlantic region, abutting the Mediterranean and Pyrenean regions along the border of Northern Spain and Southern France in seven provinces (Figure 1). The Spanish Basque Country includes the provinces of Àlava, Bizkaia, Gipuzkoa and Navarre. The Northern Basque Country in France includes the provinces of Lower Navarre,

Labourd and Soule. These seven provinces are divided into three distinct administrations; the Autonomous Basque Country (Àlava, Bizkaia and Gipuzkoa) and Navarre in Spain, and the *Departement des Pyrénées-Atlantiques* in France.

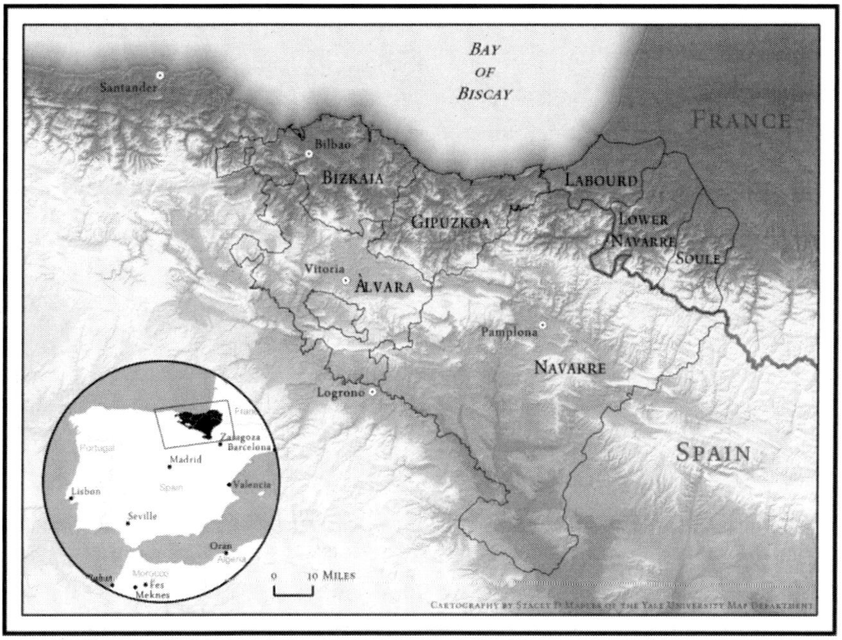

Figure 1. Map of Basque Provinces.

The total population of the Basque country is a little less than 3 million inhabitants, of which approximately 262,000 reside in the French provinces and 2.7 million in Spanish Basque country. By various estimates, the Basque population comprises between 5-6% of the total population of Spain (NSI, Eustat).

The Basque people form a culturally and linguistically distinct indigenous group in Europe with a history of approximately 30 millennia of continuous settlement in the region. Political and cultural autonomy has always been a high priority for Basque people, although they never achieved an independently governed state. The 1978 Spanish constitution defines Basque as a nationality but only Spain as a nation. Through this arrangement the Basques have their own system of governance with a two-house parliament as well as local government structures extending from provincial councils to municipal and town councils, and internally funded school systems and highways (Kurlansky 1999).

The regional climate is heavily moderated by the Bay of Biscay, with monthly average temperatures ranging from 46°F to 63°F over the course of a year. Precipitation is heavy due to the maritime and topographic influences, making the Northwest coast of the Iberian Peninsula one of the rainiest areas of Europe. In the park, precipitation reaches 79 inches a year in the valleys with as much as 110 inches of rain recorded on the mountain peaks. The mountains contain a great deal of mineral wealth, the bedrock geology being a variety of Paleozoic igneous rock, Mesozoic sedimentary rock and quaternary deposits. Veins of minerals, including goethite and siderite were exploited for the extraction of iron and galena for lead and silver since Roman times and later for iron. The legacy of these extractive industries, which remained significant until the later half of the 20th century, is still apparent in the region's mining landscapes characterized by exposed rock matrices, pools and lakes, networks of cart tracks and abandoned excavations. Despite a challenging terrain, agriculture and grazing have formed an integral component of Basque culture and traditional practices. The steep and rocky terrain led to the fragmentation of the landscape into small agricultural plots for which regionally appropriate techniques were developed, often tailored to a highly site-specific microenvironment. Traditional land use reflected application of local knowledge to imitate natural processes such as nutrient flows, hydrological systems and seasonal and altitudinal variability (Garcia-Ruiz et al 1996, Pineda 2000). This land history is still apparent in the patchwork of smallholder agriculture and the mosaic of plantation forestry plots now managed on former farmland. Historically, the land was defined as a "heterogenous mosaic of pastures and crops with hedgerows and native deciduous forests…" (Schmitz 1998). Agricultural production peaked in the late 19th century however, when shifting regional demographics and increasing industrialization and urbanization led to the decline of the rural population, and marginalization of the agricultural sector accompanied by the breakdown of its traditional social system. Abandoned cropland was increasingly reforested in plantations of exotic conifers during the 20th century. Monterey pine (*Pinus radiata* D.Don) was planted widely during the 1940s because it was a hardy and fast-growing tree. Today, 54% of the Basque country is characterized as woodland with crown coverage greater than 20% and of this forested area, 50% is Monterey pine plantations (Montiel and Galiana 2005).

Aiako Harria Natural Park was designated by the Basque Government as a Protected Natural Space, (PNS), as well as a Natura 2000 site by the European

Union. Natura 2000 is a network of 231 natural protection areas for threatened species and habitats.

Figure 2. Boundaries of Aiako Harria Natural Park and Sites Visited.

This network is a part of the Habitats Directive, a piece of legislation drafted by the European Commissions Directorate General on the Environment as part of the Biodiversity Action Plan. Its designation as a PNS indicates that it includes some of the natural formations and key ecosystems in the Basque Country that are prioritized for conservation, development and environmental education purposes (Basque Government, 2006). The park is regulated by an appointed director and a board of trustees. (Regional Council Decree 46, 1999).

Aiako Harria is located at the eastern end of Gipuzkoa province, where the territory is divided between five municipalities: Oiartzun, Irun, Errenteria, Hernani and Donostia-San Sebastian. Declared a Natural Park in 1995, it covers over 17,000 acres. The southern region of the park is predominantly mixed hardwood forests dominated by beech and oak, plantations of Red Oak (*Quercus rubra* L.), and exotic conifers such as Monterey pine and Douglas Fir (*Pseudotsuga menziesii* (Mirb.) Franco) as well as land for public grazing. The northern end of the park, which transitions into the foothills of the very western tip of the Pyrenees, is comprised of steep and forested ridges with scattered grazing.

Añarbe and Artikutza reservoirs, whose headwaters originate within the park, supply water to approximately half of Gipuzkoa's population of 673,000 people. While 80% of the forests in Gipuzkoa province are privately owned, 83.2% of the forests within the park are considered public, belonging to the municipalities. Surveys have indicated that the park is frequented by over 575,000 visitors each year, largely from the municipalities surrounding Aiako Harria (LIFE Aiako Harria).

The Euskadi Biodiversity Centre is managed by Ihobe, an environmental management public enterprise for the Basque Government's Department of Environment, Agriculture and Fisheries. The stated goal of the Centre is the conservation of nature and biodiversity and education of the public at large in such values. One of their responsibilities is overseeing the region's Natura 2000 sites.

Table 1. Management Priorities in Aiako Harria Natural Park

Management Area	Issue	Inquiry	Report Author(s)
Restoration	Create a plan to maintain those forests that are well conserved and a restoration program for those which are not.	How can *Pinus radiata* plantations be transitioned towards native forest?	Peter Caligiuri Sarah Lowry
Ecosystem Services	There are two reservoirs near the park, including the Artikutza reservoir that provides water to the city of Donostia-San Sebastian.	How can water quality and yield be better served?	Michael Blazewicz
Recreation	The park attracts a wide group of users with varied interests and needs	How do we integrate the public into the park without compromising conservation?	Abigail Adams Jason Lawhon
Legacy	The park holds some of the largest beech pollard forests in Europe.	What is the importance of preserving these forests?	Stella Cousins Ashley DuVal

A second responsibility of the Center for Biodiversity is the management of the Network of Natural Protected Spaces (NNPS), which include Protected Natural Spaces. The resource management issues implicit to the multiple use management of the park are summarized in Table 1.

METHODOLOGY

Pre-Trip Preparation

Our methodological process can be broken down into three phases. The first phase consisted of our pre-trip preparation in New Haven over the course of six weeks. We used a gap analysis to assess strengths and weaknesses of the team, identifying areas of individual and group expertise relevant to the management priorities identified by the Euskadi Biodiversity Centre. Teams of one to two students were assembled to address each of the different management themes, which we adapted to reflect what could be practically accomplished with the time and resources at hand. Our next step of pre-trip preparation involved comprehensive background research on the region, with each team member preparing a report on a different aspect of our project area and region. These included Basque political and social history; an analysis of the various organizations affecting and concerning Aiako Harria Natural Park; environmental policy and political processes affecting the region and the park; social demographics of the Basque country; and history of land use in the region. The information from this initial research was compiled into a printed report that served as guidebooks. Finally, in the week before the trip, Tamara Muruetagoiena, one of the editors of this volume and native to the Basque country, hosted a series of meetings and discussions to develop areas of inquiry and provide a briefing in the customs and social etiquette of Basque culture. The outcome of our pre-trip preparation was a broad but superficial understanding of different elements pertaining to the environment, culture and history of the region, institutional relationships and issues in resource management. Cognoscente of language barriers and possible limitations imposed by our status as foreigners, we believed our most effective contribution to be in providing an objective third-party perspective with regards to the complex political and organizational landscape.

During the second phase of this process, the official field visit, we covered a wide range of the natural features, management projects and recreational uses of Aiako Harria. In the span of one week, we met with twenty-one scientists, government employees, forest technicians, academics and policy makers as well as informally interviewed a number of park users.

Field Observations

On the first day of our visit, in the Oiartzun region of the park, we visited the riparian forest at the head of the Sarobe river, observed a forest restoration project in the valley of Karrika, and hiked around the picnic area of Oieleku to see areas of well-preserved pollarded beech trees and experimental grazing exclusion sites to promote beech regeneration. For historical perspective we visited the Arditurri mines, from which first silver and then iron had been extracted continuously for 2000 years spanning from Roman times until only twenty years ago. On the second day of our field visit in the Irun region of the park, we visited a number of EU LIFE project sites showcasing a variety of restoration projects, including artificial vernal pools to promote amphibian habitat in Enbido and a range of plantations featuring different species of exotic conifer that the Centre intends to transition back to mixed native hardwood forest. We visited the English Castle, (*Castillo del Inglés*) a popular tourist site within the park of particular importance during World War II as a part of the Comet line trail, by which allied soldiers were guided over the Pyrenees into neutral Spain to return to Britain. In Irun, we visited the Roman museum and observed marshland conservation efforts at *Parketxe Plaiaundi*, the Plaiaundi ecopark and wetland. On the third day we visited the municipalities of Errenteria and Donostia-San Sebastián. We began with a visit to the Añarbe experimental forests, where Monterey pine plantations are being converted to grassland, bioengineering techniques were applied to eroding hillsides and increased structural complexity from organic matter and standing wood was preventing erosion while enhancing nutrient cycles and riparian habitat upstream of the watershed. The day concluded with a visit to the Center of Environmental Resources in Kristina Enea Park, Donostia-San Sebastián, where we participated in a presentation and discussion with the public on the topic of forests and watershed management in the Basque Country and the USA. On the fourth day, we visited the nearby UNESCO Biosphere at Urdaibai where we hiked in Arratzu through Holly Oak (*Quercus ilex* L.) forests and some of the best-preserved forests of Pedunculate Oak (*Quercus robur* L.) in the Basque country, a characteristic feature of the poor limestone soils a drier Mediterranean climate and succession following human modification of the landscape that defines much Cantabrian vegetation.

Finally, we paid a visit to the Euskadi Biodiversity Centre and museum in Busturia and participated in a round table discussion with Centre scientists on the management of biodiversity in Basque and US forests. On the fifth day, we visited Artikutza, a private forest and reservoir servicing Donostia-San Sebastian but located in Navarre. This area has a long legacy of iron mining, and featured many lapsing stands of pollarded beeches that were managed as a local charcoal source for the mines. On several unscheduled days at the beginning and end of our planned visit, students went on guided hikes with our hosts and local experts, returned to sites of interest, and conducted unstructured interviews and participant observation with recreational park users.

Post-Trip Wrap Up

In the final phase of our assessment following the trip, we shared our notes from the field and followed up with our Basque contacts in response to new questions that arose. We conducted literature reviews on the issues we explored in our field visit and drafted recommendations that drew upon our observations and subsequent analysis. Early drafts were sent to our hosts for comments, and we conducted peer reviews of one another's work.

Summary of Papers

Adams and Lawhon's paper addresses the issue of integrating public use in the park without compromising conservation objectives. The authors conducted informal interviews with a wide variety of resource management specialists and stakeholders to ascertain values and perspectives, and map usergroups and conflicts. They conclude that the current challenges and obstacles in park governance and management include a lack of central management, unclear goals and inadequate opportunities for discussion between stakeholders. They propose recommendations for engaging the public and opening lines of communication between user groups through forums, workshops, surveys, and increased opportunities for environmental education and community service.

Blazewicz presents a series of management techniques that promote management for various ecosystem services appropriate to each of the different riparian zones; the headwaters, transitional and depositional zones.

His recommendations include adopting a "green infrastructure development approach" to work within the natural dynamics of hydrologic systems, and to increase connectivity between watersheds by restructuring road networks. This green infrastructure development approach is intended to reduce the need for ongoing management while improving ecology and water quality, reducing flood risks and improving riparian and buffer habitat.

In his analysis, Caligiuri addresses the transition of declining conifer plantations back to a mixed native hardwood forest. His analysis calls for a holistic strategy of forest management that restricts timber production to only the most productive forestland, allowing forest managers to prioritize other objectives. Drawing from the Stand Dynamics Paradigm, he calls for a "close-to-nature silviculture approach" whereby mature plantations of exotic species can facilitate a transition back to native forests through processes of ecological succession and natural regeneration.

Lowry's paper outlines the financial implications of different approaches to converting plantations to mixed species native forests in light of the downturn in timber markets and increased focus on management of forests for diversified services. She considers four scenarios, two focused on maintaining the Monterey pine plantations and two considering the transition of these systems back to mixed native hardwood forests. In each of these scenarios, the net present value of the expected future cash flows from the forest stands are calculated using stumpage prices from the literature. She concludes with an overview of the ecological impacts of different forest management approaches and their implications for ecosystem services, a perspective often overlooked in economic analyses focused on short-term costs and financial returns.

Cousins and Duval apply a historical ecology approach to evaluate the biological and cultural value of the pollarded beech and oak forests of Aiako Harria. Recognizing the time-sensitive nature of management interventions with the increasing structural vulnerability of the senescing pollards, they outline lessons from relevant case studies in the conservation and management of biocultural resources.

Pollarded stands are evaluated through their value as tools in educational, experimental stands, a point of cultural pride and value, and important contributors to biodiversity and structural complexity through the addition of dead standing wood to the landscape and recommendations are added to an overview of current projects planned under another LIFE project.

Conclusion

Rapid appraisal techniques are useful as a planning tool in resource management, and form an important component of a professional education by developing skillsets in observation and critical thinking, while encouraging interdisciplinary approaches and teamwork. Such collaborations between environmental organizations and young professionals are mutually beneficial in that they allow graduate students practical applications of their training and build professional relations in the arena of practice, while infusing novel and diverse management suggestions into programs seeking new directions or gridlocked by stalemates of bureaucratic or political nature. Increasingly, governing agencies are aware of the importance of dynamic, adaptive management of public lands in order to respond quickly to shifts in economic markets, new research and changing public values. Partnerships with foreign student groups present a cost-effective and risk-free opportunity to gain an independent appraisal built on other best-practice scenarios. While this case study reflects site-specific characteristics of Aiako Harria, it nonetheless provides a broadly applicable methodology for assessing multiple-use protected area management, particularly in Europe.

References

Basque Government (4/9/2006) Basque Country Network of Protected Natural Spaces. [Visited in October 2009] http://www.ingurumena.ejgv.euskadi.net:80/r49-4975/en/contenidos/informacion/renp/en_bio/enp.html.

Clark, T.W., Ashton, M.S. 2004. Interdisciplinary Rapid Field Appraisals. *Journal of Sustainable Forestry*, 18:2, 1-30.

Eustat(http://www.eustat.es/elementos/ele0004700/ti_Population_by_provinces_sex_and_population_density_2006/tbl0004736_i.html).

Garcia-Ruiz, J.M., T. Lasanta, P. Ruiz-Flano, L. Ortigosa, S. White, C. Gonzalez, C. Marti. 1996. Land-use changes and sustainable development in mountain areas: a case study in the Spanish Pyrenees. *Landscape Ecology*. 11(5): 267-277.

Harria. http://www.ingeba.euskalnet.net/lurralde/lurranet/lur24/erlaitz/df46.html.

Kurlansky, M. 1999 The Basque History of the World. Walker and Co., New York.

National Statistics Institute (http://www.ine.es/en).

Pineda, F.D. 2000. Intensification, rural abandonment and nature conservation in Spain. In: Examples of European agri-envrionment schemes and livestock systems and their influence on the Spanish cultural landscape. Bruce, R.G.H., M. Perez-Soba, B.S. Elberses, M.J. Prados, E. Anderson, M. Bell, P.J.A.M. Sheets, eds. *Proceedings of a European Workshop*. Montiel and Galiana 2005.

Regional Council of Gipuzkoa, Decree 46(5/4/1999) on Natural Park Management Aiako.

Schmitz, M.F., J.A. Atauri, C.L. de Pablo, P.M. de Agar, A.J. Rescia, F.D. Pineda. 1998. Changes in land use in Northern Spain: Effects of forestry management on soil conservation. *Forest Ecology and Management*. 109: 137-150.

Sundseth, K. 2005. Natura 2000 in the Atlantic region. Office for Official Publications of the European Communities, Luxembourg.

In: Multiple-Use Management ... ISBN 978-1-61942-328-2
Eds: A. DuVal, T. Muruetagoiena et al.

Chapter 1

CREATING SPACE FOR INTEGRATED MANAGEMENT OF AIAKO HARRIA NATURAL PARK, BASQUE COUNTRY

Abigail L. Adams and Jason D. Lawhon
Yale School of Forestry and Environmental Studies,
New Haven, CT, US

ABSTRACT

Management of Aiako Harria Natural Park is constrained by several factors. Fragmented ownership, governance, policies, and goals serve few people well. All participating stakeholders suffer from value deprivations, preventing management in the common interest. Although the park is attempting to incorporate new management strategies, conflicts in land management goals between the provincial government, the Basque government, and the European Union result in a stagnation of innovative ideas up and down governance levels. Conflict between municipalities, private landowners, and different park user-groups both mirrors and furthers these divisions. Analysis of the context within which policy decisions are made in the park affirm that the process does not meet standards of being comprehensive, integrated, timely, or effective. Actions must be taken to bolster efforts that manage in the common interest by creating arenas for the honest brokering of ideas and the integration of differing values. The current monodominant management paradigm needs to shift to a more encompassing, holistic, multiple-use paradigm that addresses the values of multiple user groups. Some specific recommendations include: creating a framework for participatory decision-making, including forums for education, knowledge sharing, and meaningful public feedback, increasing monitoring and appraisal of programs, and building on successful collaborative park management examples.

Keywords: Multiple-use, Policy Sciences, Management Challenges, User Conflict, Decision-Making

INTRODUCTION

Management of the Aiako Harria Natural Park in the Basque Region of Spain is shared amongst multiple entities with competing special interests. Thus challenges arise in the development of sustainable management practices that are inclusive of multiple stakeholder values. The Euskadi Biodiversity Centre defined the policy goal for our analysis as how to integrate public use into the park without compromising biodiversity conservation. In order to achieve this goal, arenas need to be open, inclusive, transparent, and viewed as legitimate by participants (Clark, 2002). Common interests are defined as shared interests whereas special interests are incompatible with comprehensive goals (Lasswell and McDougal, 1992). Many organizations, parks, and conservation initiatives are grappling with this same management challenge, and we will draw upon examples of effective value integrations in governance processes in our analysis and discussion.

Park management in Aiako Harria is primarily monodominant: one or two uses are prioritized over the broad range of potential park uses. Most recently, timber management monopolized decision-making. However, conditions and circumstances changed, shifting priorities towards biodiversity conservation. Transition towards a more integrated, multiple-use management paradigm is complicated by biophysical, social, and decision process challenges. These include: multiple layers of governance, a lack of effective communication and collaboration, conflicting interests of user groups, entrenchments of perspectives, and a history of technocratic land policy (Perez and Groom, 2000). An alternative to the status quo would be a practice-based, contextual, and holistic management paradigm. Specifically, this would involve integrating stakeholders, opening lines of communication among user groups, increasing transparency of actions, incorporating value rankings into the decision-making process, increasing monitoring and appraisal of programs, creating benchmarks of success, and creating a protocol for the termination of ineffective programs.

This analysis will define the policy problem that Aiako Harria Natural Park managers face, analyze the trends and conditions that have led to this circumstance, project future trends, and offer recommendations for the integration of values shared between citizens and managers in the common interest.

METHODS AND STANDPOINT

We conducted our rapid assessment of the Aiako Harria Natural Park from October 17-23, 2009. Prior to the trip, the class split up the tasks of researching background information, conducting a rapid stakeholder analysis, and mapping the biophysical, social, and organizational context. We traced the political and social history of the Basque Country, mapped organizational structure, culture, and interrelationships, clarified environmental laws and policies, assessed social demographics, researched trends in agriculture and forestry, and delineated the biogeography and geophysical elements of the park.

Upon arrival in the Basque Country, we informally interviewed a variety of resource management specialists established through our hosts, the Euskadi Biodiversity Center. The center's stated purpose is to encourage, generate, and manage knowledge, both scientific and technical, for improved policy making, and to increase awareness of the value of biodiversity (Euskadi Biodiversity Centre, 2009). The center's mission is to illustrate the variety, function, and richness of our environmental capital through educating society on the roles, services, and values that biodiversity provides (Euskadi Biodiversity Centre, 2009). We met with biologists, hikers, past and present park managers, and environmental technicians from the municipalities that surround the park. Conspicuously missing from our interview pool were park user groups with differing perspectives from the Euskadi Biodiversity Centre such as hunters, farmers, herders, private landowners, and foresters. We did our best to ascertain their values and perspectives through limited interactions, but acknowledge that we did not fairly get their opinions. After returning from our trip, we reviewed our notes, debated and discussed our assessment, and conducted literature searches on the policy process, effective governance paradigms, and case studies of successful multiple-use practices in other parks.

The goals of participants in any policy process are not only a reflection of their values, but also their standpoint or perspective on the world (Clark, 2002). Our view of how the world operates, or should operate, depends on our personal and professional experience and organizational norms. Because our perspective inherently affects how we understand, frame and analyze the world; it is important to clarify our standpoint as well as the methods we used to collect our data.

We are both masters' candidates at the Yale School of Forestry and Environmental Studies and thus recognize an inherent bias towards viewing conservation challenges through our training in ecological sciences and protected areas management. We also recognize that our affiliation with Yale University may affect how we are perceived by those we interviewed. Although our backgrounds differ somewhat, we are both from the western United States where conservation challenges tend to be polemic and incite impassioned responses from citizens. Combined, we have fifteen years experience working for the US National Forest Service and the US National Park Service, with experience in wilderness, rangeland, and multiple-use management. Between our work experience, education, and training, we have been exposed to multiple cases of balancing wildlife and biodiversity conservation with human recreation, accessibility, development, and economic growth. We believe the lessons we have learned are applicable around the world in similar conservation challenges. As outsiders approaching this analysis, we tried to be as objective to all perspectives as possible. However, the rapid nature of the assessment restricted our exposure to only a limited subset of stakeholders. All recommendations we suggest are made through our best efforts to assess the situation to the best of our abilities with the goal of clarifying and securing the common interest.

POLICY PROBLEM

The organizations co-managing Aiako Harria represent various participants pursuing both compatible and competing goals. Mapping these participants, their perspectives, strategies, and desired outcomes will better illuminate the problem and provide insight into potential alternatives to manage in the common interest. The common interest is an interest that is widely shared within a community of stakeholders and is demanded on behalf of the whole community (Clark, 2002). In contrast, special interests generally serve few people at the expense of the whole community (Lasswell and McDougal, 1992; Brunner et al, 2005).

Problem Orientation

To be most effective in decision-making, one must fully understand the problem at hand before proposing solutions (Clark, 2002). Problems are socially construed, context-dependent, and heavily laden with values and expectations. We perceive a problem when a discrepancy occurs between our expectation of what will happen and the reality of what actually occurs. Too often, problems are defined narrowly and reduced to technical or scientific challenges centered around biophysical entities. In reality, conservation challenges are multi-faceted, complex, and inclusive of social interactions and decision-processes. *Problem orientation* can be used to understand and comprehensively address these issues (Clark, 2002). The five tasks of systematic problem orientation include clarifying goals (values sought), describing trends (reviewing relevant history), analyzing conditions (explanation of why trends might exist), projecting future trends, and suggesting alternatives (Clark, 2002). Furthermore, to understand and frame the problem, one must map the social and decision process context within which a problem occurs. Both of these processes will be described further in the paper after we establish the context of the situation.

Context

The Aiako Harria Natural Park is located in northern Spain, on the southwestern border of France. Designated as a Natural Reserve in 1995, the park encompasses approximately 6700 hectares, approximately 80% of which is publicly owned between five municipalities. The remaining area is privately owned, with a small portion (1%) owned by the Gipuzkoa provincial government (Life Aiako Harria, 2009). The Basque region of Spain surrounding the park is relatively rural with five main industrial towns: San Sebastian-Donostia, Irun, Oiartzun, Errenteria, and Hernani. Based on interviews and observations, the park is a very popular destination for various types of recreational activities. Additionally, forests are vitally important and have been utilized by generations of Basques to support their livelihoods.

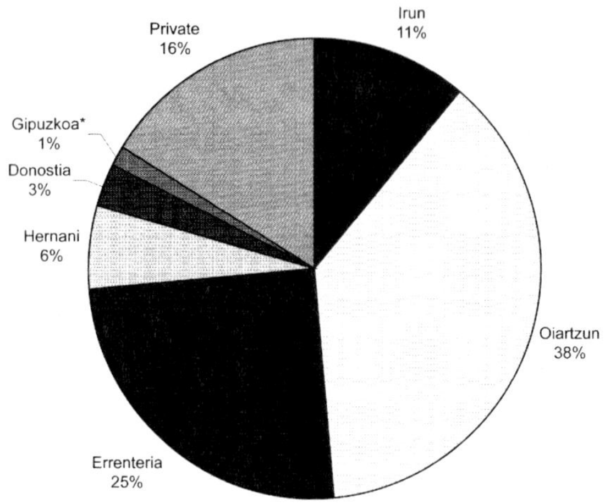

Figure 1. Distribution of Ownership of Aiako Harria Natural Park.

Social Process

The arena where people shape, share, and promote their values to advance their interests is known as the social process. In order to define this process, the situation, the participants, their perspectives, their values and strategies of action, and the outcomes of all these elements, are mapped (Lasswell, 1970). A grounding theory in the policy sciences is that humans seek values through institutions using and affecting resources (Lasswell, 1971). Deprivations of human values usually lie at the heart of a problem rather than technical, scientific, or logistical challenges that are often relied upon to explain a failure in the system. Thus, if value outcomes are to be in the common interest, the social process arena needs to be open, inclusive, transparent, and viewed as legitimate by all participants (Clark, 2009).

Conversations with a member of the Gipuzkoa Mountaineer Federation and an aquatic ecology professor from the University of the Basque Country suggest a growing mindset shift in the urban, industrialized Basque cities. They speculate that compared to a historic philosophy of accessing mountains for resource extraction and subsistence, urban users today desire a park for hiking, recreation, and biodiversity conservation. Private landowners may differ in values from urban populations due to their different connection to the land through practices of livestock grazing, timber harvesting, hunting, and the extraction of non-timber forest products such as mushroom harvesting, though the latter two activities occur irrespective of land ownership. The provincial government of Gizpukoa also has a more traditional perspective on resource use for the park, including a continued focus on timber management. This may be due to their reflection of an older generation's values.

This discrepancy in goals between the Basque Government, the provincial government, municipalities, and private landowners is magnified since ownership of the park is divided between these entities. Furthermore, the former park director explained that the Basque Government hasn't clearly defined goals for Aiako Harria. A lack of central management of the park, unclear goals, and inadequate arenas for discussion have all limited transparent and constructive dialogue to express tensions and value divergences. This results in fragmented attempts at coordination. In addition, competing values between hunters and hikers, private landowners and urbanites, foresters and biologists, and managers at various levels of government, lead to conflict and confusion over park goals.

The problem, as defined by the Euskadi Biodiversity Center, is that they are trying to fit recreation needs of multiple users and the demands from timber interests, private landowners, foresters, and biologists alike into their biodiversity conservation goals. Our perspective on the problem is that a history of monodominant management, fragmented ownership, and a focus on engineered and technological solutions, has led to governance that is not in the common interest.

The tests of the common interest include a *procedural test* to determine if decision making is inclusive, participatory, and representative, a *substantive test* that asks if concerns are valid, appropriate, and broadly supported, and finally *a pragmatic test* that determines if participant's expectations have been upheld such that policies work in practice (Cromley, 2002). The Euskadi Biodiversity Centre seems to rely on scientific knowledge to guide decision-making and education to inform the people about biodiversity. This perspective is in many ways as single-use focused as previous paradigms for the park have been. Aiako Harria lacks a coherent, integrated, multi-method arena for open-dialogue and honest brokering. Interest groups are not equally represented and do not feel that their concerns are validated. From conversations with individuals, the most commonly expressed denied values included well-being, respect, and rectitude. When individuals feel that they are not being respected and that their well-being is threatened, they may become hardened in their attitudes, creating an "us versus them" mentality among participants.

ORGANIZATIONS

The conservation initiatives backed by certain park users often begin at the European Union level. Out of the Pan-European Strategy for Biological and Landscape Diversity emerged the EU's Biodiversity Action Plan (BAP, 2008). From this strategy emerged the Natura 2000 plan with its dedicated purpose to increase biodiversity conservation through a European ecological network of protected areas (Council Directive 92/43/EEC, 1992). These habitat directives are then sent through the Basque and Gipuzkoa provincial governments for implementation. The provincial government, unlike the EU, seems more intent on maintaining a more utilitarian focus in Aiako Harria. This assumption is based on conversations and from their stated priorities of land management and improvement of forest resources, livestock, conservation of nature, and the protection and promotion of hunting and fishing (Gizpukoa provincial government, 2009). The Gizpukoa provincial government also oversees management of the park and the board of trustees who are chosen to represent the differing interests of park users. Additionally, many of the management plans and protection measures developed for the park are actually developed by private consulting companies contracted by the provincial government, such as Nekazal Ikerketa eta Teknologia (IKT).

Ownership and management fragmentation continues with the municipalities who each own separate parcels of the park and may have differing goals as well. Irun is the most industrialized of the municipalities surrounding the park, but is moving from a mindset of timber extraction to a greater realization of recreational and multiple-use values of biodiversity conservation and cultural maintenance (Irun Town Council, 2009). The town works to improve trails, to educate the public through interpretive signs and an information center, to maintain livestock grazing in a controlled manner, and to create forums or *quality circles* to actively involve citizen feedback (Irun Town Council, 2009).

The town of Errenteria is also attempting to implement strategic action plans to improve sustainability. As part of this effort, Errenteria is eliciting feedback from citizens through Quality of Life surveys (Errenteria Town Council, 2009). The town is beginning to recognize the benefits of social outreach and co-responsibility in framing town policies that affect all citizens. Finally, the rural town of Oiartzun has been working to convert single-species stands to mixed species forests. A clear goal of the town council is to restore and re-grow native habitats. They have worked to reforest 400 ha of town-owned properties and have acquired an additional 120 ha through land swaps.

Much of Oiartzun's land protection practices result from their policy of using their own water resources rather than joining the Añarbe Water Company along with other municipalities in large water boards.

Other Participants and Stakeholders

Participants in addition to organizations and government bureaucrats who have a stake in the park and may or may not be included in the policy process include: hunters, hikers and other recreational users, mushroom pickers, foresters, environmentalists, private landowners, and park managers. The perspectives of these different groups vary substantially. Hikers and other recreationalists including climbers, mountain bikers, and picnic day-users, tend to be aligned with some of the values of biologists, and some of the park managers in that they desire that the park be protected for recreational activities. However, opinions are nuanced between recreationalists. For instance, off-road vehicle users differ from hikers in that the latter may prefer fewer roads and more "natural" aesthetic forest qualities for walking. Climbers may become upset when a wall is close off to protect a nest and picnickers may be disgruntled when they are prevented from driving their cars to core areas of the park. Furthermore, from conversations with a few local residents and biologists, some recreational users may desire a park with "clean" forests they define as evenly spaced trees clear of deadwood, underbrush, and understory complexity.

In contrast, biologists in the park are working to transition single-species stands to native forest and to increase structural complexity for increased biodiversity. These various biologists, environmental technicians, and park rangers value measures to further sanction and protect biodiversity. Thus, park managers are left to balance two competing views of an ideal forest between these users while simultaneously addressing numerous other viewpoints and constituents.

Currently, park biologists and environmental technicians frame foresters, some private landowners, forest engineers, and to an extent, hunters, as in opposition to their goals of increased conservation measures. Forest engineers in the provincial government of Gipuzkoa are perceived as preventing conservation from going forward with their view to maximize monetary benefits from the forests and to rely on engineered projects as solutions to environmental problems. For different reasons, hunters are viewed as not being in agreement with many of the conservation goals. Like a few other recreational users, a hunter we talked with lamented the *sucio* or "dirty" condition of the forests today, inhibiting his ease of movement throughout the park. He felt that management for environmental objectives had resulted in degraded plantations and increased understory growth and woody debris in the forests. Conversely, hunters of woodcock and wild boar prefer closed forests with dense undergrowth and brambles. [Figure 2, Figure 3]

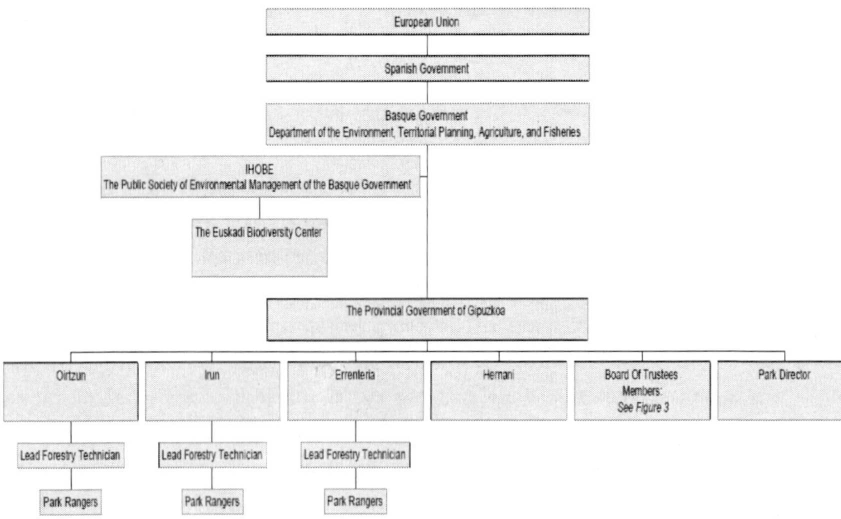

Figure 2 The different management tiers of Aiako Harria Natural Park. The municipalities of Hernani and San Sebastian also own small sections of the park, as seen in Figure 1, and have representation on the Board of Trustees (see Figure 3).

High numbers of recreational users in a relatively small park exacerbates the conflict between hunters and hikers in Aiako Harria. During the warmer months, hikers, bicycle riders, and mushroom collectors pack the park all days of the week and especially on the weekends. By local estimates, on any given Sunday, you might find 300 mushroom collectors, 100 mountain bikers, and even more hikers in the park, leading to substantial conflict between hunters and hiker/recreation groups.

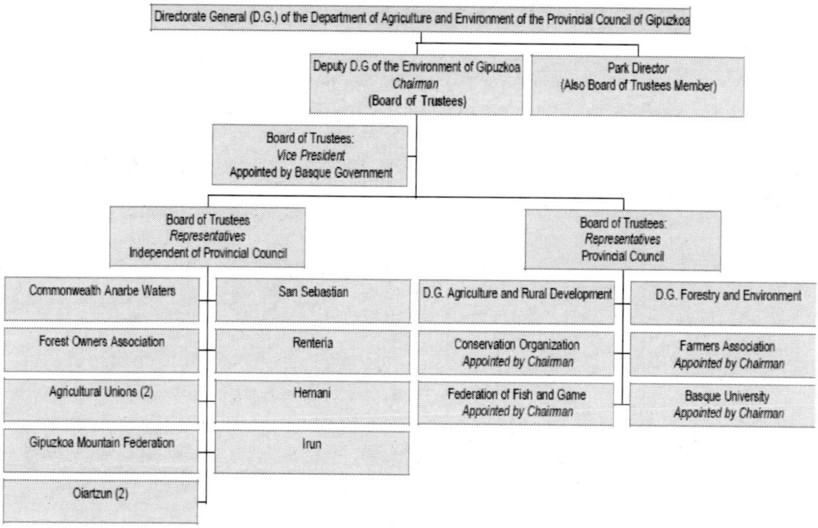

Figure 3. Management of Aiako Harria Natural Park.

Much of this may be due to the proximity of the hunting blinds in relation to the trails and designated picnic areas. It was very common to be hiking on a major trail within 20 meters of a hunting blind. At one point, we even visited a children's park where the playground was directly across a small field with numerous hunting blinds. There are approximately 500 designated hunting blinds throughout the park for hunting birds; additionally, it is estimated that 300 wild boars are harvested each year. Whether or not perceived safety concerns between hunters and recreationalists are real, people desire to use the same limited space for different, potentially incompatible activities.

Finally, conflict exists between private landowners in the park who graze livestock in common lands and practice agriculture, and hikers and recreationalists from urban municipalities who may hold conservation values. Promoters of biodiversity conservation lament the grazing of sheep, horses, and cows on the understory regeneration of forests. Furthermore, erosion problems resulting from livestock traversing over steep slopes exacerbates this conflict. Meanwhile, shepherds promote the closing of certain roads as they perceive recreationalist's dogs as a disturbance to their flock. These concerns echo the overarching conflict between an older generation that tends to value traditional and utilitarian uses of the park versus a younger, more urban generation that values conservation and recreational activities.

Table 1. Trends In Aiako Harria Park Management

Trend	Interpretation	Projected Future	Potential Future
Biodiversity Loss	Historic land use, utilitarian doctrine, inadequate monitoring	Habitat improvement, maintained/increased biodiversity but at the cost of polarizing other stakeholders	Increased monitoring, transparency, and education could lead to increased project support from other stakeholders. Biodiversity conservation would not need to be sacrificed in order to work toward common goals.
Lack of idea/value exchange between upper levels of government and municipalityes and local stakeholders	Provincial government values and goals are not aligned with the EU or the Basque Government and are therefore not being given consideration. Loss of investment due to declining revenues from timber.	Limited flow of information, values, and ideas up or down government levels. Progress forward will be difficult.	The Provincial government could play a key role as an integrator finding creative interdisciplinary goals to unite practical bottom-up objectives with well-supported landscape/regional level top-down priorities.
Trail Erosion/ Disrepair	Divided ownership, heavy use, low priority, contracting projects	Lack of maintenance and increased use will lead to further trail braiding and erosion.	Unification of land management objectives including a holistic trail inventory and prioritization could result in better, more focused management. Additionally, establishing some

Trend	Interpretation	Projected Future	Potential Future
			sort of community involvement "work days" including both general recreation users and hunters, and mushroom pickers would be in the common interest.
Hunter Dissatisfaction	Hunting areas restricted, change in vegetation characteristics ("sucio" forests), lack of communication	Hunters represent a major user group in the park. Lack of inclusion of these values could result in increased polarity and continued conflict with this user group.	Creating an arena where hunters can learn more, not be told, about habitats and biodiversity could reap benefits for all stakeholders. Recognizing their connection to the land is valid and equal to that of a scientist or land manager may subside user conflict.
LIFE Projects receiving priority	Projects are selected and funded based on EU priorities.	This top-down approach does not adequately address stakeholder values and concerns at most levels. While these projects may continue to be viewed as successful in terms of EU biodiversity conservation objectives, lack of inclusion will likely result in continued	EU funding and leadership on major issues can play a critical role in this process. However, priority coordination with local resources, stakeholders, and other levels of government is in the best interests of the park and could go further to achieving

Table 1. (Continued)

Trend	Interpretation	Projected Future	Potential Future
		conflict between user groups and dissatisfaction.	biodiversity objectives.
Prototype Municipality Projects	Projects are implemented on the ground based on local manager and municipality priorities.	Tendency to get a "bubble" effect where habitats and issues that cross boundaries are not taken into consideration.	Idea exchange combining top down and bottom up strategies has potential to result in increased efficacy of management
Timber Harvesting (foresters)	Low timber values and increased global competition has halted timber harvesting and reduced the maintenance of stands.	Stands will continue to degrade, reducing the feasibility of future harvests. However, stands left alone could result in increased biodiversity. Yet this dense undergrowth may not be favorable to hunters.	Increased maintenance of higher value stands in combination with limiting grazing to allow for natural regeneration gaps could result in areas of "better" hunting and increased biodiversity.
Lack of Public Involvement	Lack of unity in management goals, top down approaches, conflict at different levels of government, lack of arenas for full participation	The public will continue to be apathetic or react negatively if their values are not considered in the management process.	Increased public involvement is essential to manage Aiako Harria in the common interest.

Trend	Interpretation	Projected Future	Potential Future
Loss of Quality Grazing Lands	Halting of prescribed burning and maintenance, shift in population values	Decline of pasture maintenance has resulted in increased grazing pressure on forested stands. Grazing pressures have reduced natural regeneration and increased biodiversity loss.	Maintaining higher quality grazing lands has potential to better meet demands with less impact to biodiversity. Additionally, maintaining vegetation at different phases of succession could result in niches for certain grasses and forbs.

In the social process, participants utilize different strategies to shape government resources to acquire desired values of well-being, wealth, respect, rectitude, affection, skill, power, and enlightenment (Clark, 2002). Delineating parks with management boundaries is an entirely socially constructed process (Clark, 2002). It is important to recognize that ecosystems sustain all life and thus we need to promote ecosystem function, structure, and resilience (Walker et al., 2004). In addition to attending to biophysical elements, we need to concern ourselves with what human resources we are trying to conserve, including social capital, sustainable economies, and cultural components (Clark, 2002). Many park users are members of hiking and hunting associations that work to advance their special interests. However, perceived power imbalances between these two groups can lead to value deprivations, anger and resentment, and outcomes that are not in the common interest. Although all park users are unlikely to agree on a single, unified goal for park management, it is likely that common ground may exist in groups desiring that value integration and management practices should improve. [Table 1]

In addition to the social process, all problems take place in a decision arena where information is gathered and policies are promoted, debated, selected, and implemented. Brewer describes problems as having a life where they are developed, framed, estimated, and alternatives crafted, to attempt to resolve the issue.

Decision Process

This life cycle of a problem is known as the *decision process* (Brewer, 1983). Mapping this process will help to clarify the problem and frame alternatives. Ideally, the decision process as a whole should be dependable, comprehensive, and integrative (Clark, 2002).

Intelligence

The initiation or beginning stage of the decision process occurs when a problem is recognized, clarified, and estimated (Clark, 2002). Both scientific and social intelligence is gathered to help inform a later decision. Often, park managers will rely heavily on biophysical data, concluding that science will provide the right solution to the problem, regardless of the social context. In Aiako Harria, everyone we conversed with felt a particular value deprivation, suggesting that values were not integrated effectively and solutions not made in the common interest. Many of the people informally interviewed stated that park management was fragmented and split between many levels of government, with limited communication and collaboration. Intelligence gathering appears to be limited and ad hoc. There is not a strong history of scientific research in the park and the monitoring of experimental plots to date is quite restricted.

Furthermore, beyond biophysical research, social data on participants, their perspectives, needs, and goals for the park is limited in scope. Information that has been gathered largely stems from forest engineers who were trained in a technical forestry school or from the European Union LIFE project (Perez and Groome, 2000). Thus, it is unclear that a thorough and analytical estimation of the problem has actually occurred. Standards of a dependable, comprehensive, and transparent process have not been met. Additionally, due to varying perspectives and values, it is unlikely that user groups will clarify a single and unified goal for the park, furthering a shared sentiment of fragmentation and disorganization.

PROMOTION

The next stage in the decision process is the promotion or open debate phase. This is where participants dedicate themselves to finding solutions to the problem as they understand it (Clark, 2002). Many groups promote their viewpoint through media outlets, lectures, and publications. Often, a conclusion is made that more research is needed to reduce uncertainty in the decision process. The Euskadi Biodiversity Centre is enveloped in the promotion phase through their emphasis on educating the public, creating displays on biodiversity, and promoting their values through brochures, publications, and mailings. Furthermore, this singular goal is emphasized over other values such as sustainable timber harvests in appropriate areas, investments in cultural preservation, grazing rights of private landowners, or the promotion of hunting and recreational use of the park due to funding from EU LIFE projects.

The provincial government established a board of trustees to facilitate promotion and to ensure participation and representation of different interest groups in the management decisions of the park. The director keeper of the board is the head of management and administration of the park, reporting to the Directorate General of Forestry and the Natural Environment (See Figure 3). Although the formation of this board moves in the right direction for facilitating integration, many of the board members are appointed by the provincial government and not all interests are fairly represented or validated. Thus, the current promotion arena does not meet the standards of being rational, integrative, or comprehensive. This leads to policy options that are not in the common interest.

Efforts are beginning at the Euskadi Biodiversity Centre and elsewhere to facilitate dialogue between user groups, create forums for discussion, and to survey individuals to get a better sense of value discrepancies and expectations. Encouraging such arenas where trust, honesty, and open debate are valued will allow space for more inclusive perspectives.

SELECTION, ENFORCEMENT, IMPLEMENTATION

After innovative alternatives are suggested and promoted, the weighting, choosing, and application of a policy form the selection, enforcement, and implementation stages of decision-making. When selecting an alternative to apply, managers develop the rules and norms of conduct for how the policy will be implemented (Clark, 2002). Ideally, this process allows for open debate between different interests to determine where common ground and consensus exists. If small groups can work together to build trust, it is more likely they will find common goals between them that they can work toward implementing.

Currently in Aiako Harria, it appears that hunter associations have significantly more political power and money than recreational hiking and mountaineering associations. Of the approximately 2 million people in the Basque Region, there are approximately 100,000 in hiking and mountaineering associations and 20,000 in hunting associations. The hunters group form a strong lobby using diplomatic, ideological, and intimidation strategies to afford respect for their position.

While striving for the values of respect and power, they are depriving others of the values of respect and well-being. These value deprivations suggest that selection standards of effectiveness, comprehensiveness, and integration are not being met (See Table 2).

Furthermore, although management decisions are affected every four to five years by changing political leaders and park managers, it is important to establish a forum where central goals and themes are clarified and consistently carried through to fruition. During our visit, we learned that recently crafted biodiversity strategies and public-use documents for the park were stalled by request of the new park manager. Implementation of these documents has not occurred. The current situation induces a bottleneck at the provincial government where current and future projects are frozen. Innovative and new ideas cannot move up or down through the decision process in this structure.

This evidence suggests that policies are not implemented in a timely or dependable manner, falling short of those standards. Moreover, it appears that most of the environmental directives come down from the European Union and are then interpreted on the ground through the implementation process. In one informal discussion with a local manager, it was explained that the law tends to be loosely interpreted in Spain.

Table 2. Social Processes in Aiako Harria

Participants	Perspectives/Goals	Situations	Base Values	Strategies	Outcomes	Effects
Park Management	Manage Park Effectively	Participate through direct observation and interaction with users	Power, Skill, Respect	Educational Economic	Small prototype projects with individual goals.	Small steps toward individual goals but little monitoring or movement towards managing for the common interest
European Union	Stop the decline of endangered species and habitats in the EU by 2010	Participate through establishing directives, funding projects, some monitoring	Power, Enlighten ment, wealth	Diplomatic Economic	Biodiversity Action Plan, Habitats Directive, Water Directive	Stop biodiversity loss at the expense of other participant values
Basque Government	Restore native species and increase biodiversity	Write EU directives into law. Participate through Ihobe and the Biodiversity Center	Englighten ment, Skill, Respect	Educational Diplomatic	Alignment with E.U. goals results in polarizing Provincial Government	Little knowledge and value exchange up or down
Provincial Government of Gipuzkoa	Maintain livestock, improve forest resources, conserve nature, hunting and fishing	Oversee park management	Respect, Power, Rectitude	Diplomatic Ideological	Values and concerns are not given legitimacy resulting in a stalemate	Even projects with characteristics favorable to the common interest have no chance of implementation

Table 2. (Continued)

Participants	Perspectives/Goals	Situations	Base Values	Strategies	Outcomes	Effects
Municipal Governments	Manage parks, gardens, and green space within municipal boundaries; including their portion of the park	Participate through direct observation and interaction with users	Power, Skill, Respect, Well being	Educational Ideological Diplomatic	Management goals based on municipality boundaries. Minimal incentives to work across boundaries	Fragmented park management. Degradation of resources. Park not managed for the common interest
Hikers, bikers, climbers, picnickers, etc	To enjoy and recreate in a natural area "wilderness" that remains accessible and consistent	Unrestricted access, scenic vistas, maintained trails and facilities	Well-being, Rectitude	Diplomatic	Recreationalists often in opposition to hunter groups over shared space/value differences	Opportunities to find common ground are missed because of historic/current conflict and the park as a whole suffers

Laws are viewed more as guidelines than as hard and fast rules. He made the analogy that speed limits in Spain are viewed as loose suggestions around a general limit, much the same way that rules and regulations are interpreted. His point was that unless there is an advocate for an environmental cause, directives from the European Union will not be implemented as they were imagined to be. Again, this evidence suggests that standards of effectiveness, dependability, and rationality are not being fully met (See Table 1.)

It also appears that in policy implementation, groups are often hardened in their attitudes. It was apparent that different perspectives felt that the other "thinks they know best" and that "they don't listen to us." As one park manager put it, "the only way to get things done around here is through fighting." Yet, a lack of participatory approaches prevents trust, listening, and clarification of the common interest. The same manager made the comment that integration of participatory approaches was very weak in the Basque Country. When we asked a park-affiliated biologist about how the public viewed management practices in the park, he jokingly responded, "What public? We don't have any people." Such barriers of information and entrenched disciplines prevent adaptive and sound governance measures from effectively moving forward. In response, enforcement of park rules becomes difficult. For instance, a few weeks before we arrived, a motion was passed to remove six of the approximately 500 hunting blinds in the park due to safety concerns. After removal, 80% of the interpretive signs in the park were cut down and graffiti appeared with the name of the lead environmental technician of the local municipality in the center of a bull's-eye. This power struggle mirrors what is happening between the provincial government, the public, and the other levels of government.

Evidence suggests that policies are not meeting standards of being integrative, comprehensive, constructive, and provocative (See Table 1). If tensions are high enough that violence and anger become the outlet for value expressions, the common interest is insufficiently served through the status quo.

Appraisal and Termination

The final stages of the decision process include the appraisal and termination functions where policies are monitored, evaluated for effectiveness, and terminated if they are not successful (Clark, 2002). From discussions with non-profit organizations and park managers for Aiako Harria, there does not appear to be a formal method for how projects are formulated, monitored, and evaluated for effectiveness. Arguably, this function may be the most important in the decision process as it allows progress to be benchmarked against goals. Clarifying whether a particular practice or policy is really working and serving the purpose for which it was intended is essential for moving closer to the common interest. If, however, it is found that a policy is not promoting greater unity or bestowing benefits, then it should be terminated in a timely and efficient manner. This allows room for more effective alternatives to be shaped, discussed, and implemented. In Aiako Harria, it is not evident that ineffective policies are terminated in a justified and timely manner (See Table 1). Individuals and organizations on both sides of an issue have no means of enforcement or appraisal. Regulations coming from the EU are often taken relatively lightly. Divided management, poor appraisal, and limited termination in the decision process leads to non-accountability and few stakeholders being served by programs.

Analysis of the Problem

To be fully contextual in mapping, analyzing, and understanding the problem, it is useful to trace historic trends, explain why those trends or conditions exist, and make projections for what will likely occur in the future if the status quo is continued. Often, problems are narrowly defined or misconstrued due to a failure to account for past trends, explanatory conditions, future projections, and value dynamics (Clark, 2002). We will analyze biophysical, social process, and decision process trends, conditions, and projections to clarify our analysis. Finally, we will summarize the problem as we perceive it.

Trends

Biophysical trends in Aiako Harria include forest biodiversity loss compared to "control" forest sites that haven't been managed in the last 100 years, ridge-line trail erosion, degraded and over-crowded plantations, marginalized grazing land, slowly decaying beech pollard forests, increased dirt track coverage, and increasingly aggressive forestry techniques. Social process trends include a decline in community volunteerism, a decline in a subsistence agriculture and traditional practices, an increase in industrialism, a shift in population values, and a disconnect in communication, values, and integration of ideas across user groups and levels of government. Decision process trends include a lack of transparent exchange of values, ideas, and stakeholder perspectives in an integrative and comprehensive manner, a lack of public involvement in decision-making, a fragmented and non-unified management system, an imbalance in user-group political clout, and a disconnect between goals and objectives across various levels of government (See Tables 1 and 2).

Conditions

The reason the biophysical trends discussed above exist is due to many factors. The decrease in "biodiversity" (as defined by an index that accounts for species diversity and abundance, is likely due to an extended history of over 500 years of human manipulation and management of the environment. Originally this management existed for subsistence and agriculture purposes, but shifted to a heavy focus on timber harvests and plantations in the 1950's as people began working in the factories (Perez and Groome, 2000). In addition, a lack of structured monitoring of forest quality has prevented a precise and clear understanding of forest trends and conditions to be accurately compiled (See Table 1).

The 1970's marked the end of Francisco Franco's regime in Spain, inducing many cultural, land-use, and lifestyle shifts in the Basque Country. Along with these cultural changes, the end of the Franco dictatorship opened up democratic rule that freed up trade with other countries (Perez and Groome 2000). People desired greater economic gains in the form of timber production and exports. Monterey pine *(Pinus radiata* D.Don*)* plantations dominated the landscape at this time after a 1956 frost killed most of the pines in Gipuzkoa. People migrated to burgeoning cities and the more traditional subsistence lifestyle declined. During this time, management of this area seemed focused on a single purpose: the planting, growing, and harvesting of Monterey pine largely for paper mills. Today, due to current low timber prices and competition with France, Chile, Russia, and New Zealand, Monterey pine that was planted as a cash crop for the municipalities, provincial government, and private lands alike, is relatively worthless. It costs more to remove the timber than it is worth and loggers have agreed not to bid on timber sales at current levels. Along with various pests and illnesses, this has been one contributing factor towards the deterioration of the pine stands throughout the park.

These explanatory conditions are linked to the social process trends discussed above that arose through changes in land management and values post-Franco. In an interview, we learned that up until thirty years ago, community volunteerism was common in the area. Neighbors would come together to do trail maintenance, and farmers, sheepherders, and foresters would work together to conduct regular prescribed burns for the maintenance and improvement of rangeland. Subsistence farming was a way of life. One such practice that has become increasingly rare is the use of *metas* by piling bracken fern to be used as bedding for livestock in the winter, then applied to the fields as fertilizer. Prescribed burns were used to clear pasture for livestock and fern growth. The decline of this practice at farms results in pasture conversion to forests and a reduction in suitable grazing land. With a lack of highly productive grazing land, livestock graze in more marginal areas, further halting regeneration in forests (See Table 1). However, livestock have grazed in the forests for more than 200 years, halting natural regeneration for centuries.

Decision Process trends discussed above can be explained by governance challenges and disjointed goals. Today, controversy swirls over "utilization" versus "conservation" values and best management practices between decision makers, user groups, and individuals. The Euskadi Biodiversity Centre has recognized this cumulative shift in how nature is perceived as an opportunity to change management strategies in the park. However, it seems that their goal is still predominantly single-purpose: to shift from timber production and traditional uses to biodiversity conservation. Although it is easy to understand why the latter goal needs to be emphasized when the former uses are driven by economics, a monodominant use paradigm invalidates the values of those not included in the process. The Biodiversity Center's goals are backed by the European Union who fund and support projects aligned with the LIFE initiative to increase biodiversity conservation. As stated on the LIFE website "LIFE is the EU's financial instrument supporting environmental and nature conservation projects throughout the EU..." (BAP, 2008). Additionally, these goals are also consistent with the goals of the Basque government. However, the provincial government of Gipuzkoa has maintained a history of valuing utilitarian uses for the park. Because these goals are often in opposition to the conservation objectives of the Basque government and the European Union, it is likely that the higher governance levels do not legitimize and respect most of the provincial government's values (See Table 1). These deprivations are probably felt more intensely today due to a sharp decrease in potential timber revenues that once might have been sought as a source of power and influence for the provincial government. Today, most project funding seems to come predominantly from the European Union, prioritizing biodiversity projects aligned to those goals (See Table 1).

Further adding to a disconnected and incomprehensive decision process is the lack of central management of the park outside of the guiding direction of the park director and the board of trustees. The board of trustees' goal is "to ensure participation and representation of different interests group in the management activities of the park" (Regional Council Decree 46, 1999). However, nearly half of the board are members of, or appointed by, the provincial government (See Figure 3).

An example of the confusion of divided management and organization of the park lies in the conditions of the trails. Trails are braided and heavily eroded in areas of steep mountainous terrain (See Table 1). There is also an abundance of *social trails*, unofficial paths eroded by foot traffic, despite excellent markings and signage. Trail maintenance is contracted out to private and public consultants, and the municipalities have no mandate to work with one another. Therefore, one town may decide to invest money fixing a trail, but that goal may not be carried across ownership boundaries. Thus, conditions that lead to apathy and disrepair may prevail. However, the town of Irun accomplished a positive trail restoration project six years ago, rerouting a popular summit trail and restoring a highly eroded area.

Future Trends

Despite disjointed governance, value deprivations, and fragmented leadership, efforts are being made to increase collaboration between political levels, to work across ownership boundaries, and to increase public education and transparency. However, on-the-ground implementation and enforcement of these efforts is limited. The Euskadi Biodiversity Center is working to begin incorporating social science research along with its natural science research for management plans and policy suggestions. Additionally, the educational and public participation components of their organization are also an encouraging step in this direction.

An example of these efforts to work across governance levels to accomplish goals lies in the management of the Anarbe reservoir. A neighboring province owns large portions of this watershed, and thus management of the reservoir must be coordinated between multiple municipalities since it provides the main source of drinking water for San Sebastian-Donostia and the surrounding municipalities other than Oiartzun.

Before the watershed was protected, it was a popular recreation destination for picnicking, hiking, mushroom gathering, and four-wheel vehicle use. Thus, environmental technicians from the surrounding towns had to coordinate action steps to mutually advance both individual and collective interests, while balancing recreation demands of park users. From informal conversations, we learned that the municipalities coordinated in restoring a degraded road, constructing a wetland for amphibian habitat and educational instruction, setting aside restricted-use areas for explicit watershed protection, and conducting a scientific research study on the positive effects of coarse woody debris in streams for habitat and sediment retention while improving fishing opportunities. Although stakeholder involvement in the decision process could have been improved, the reservoir serves as an example of encouraging movement towards a more integrated governance model.

Although examples like the Anarbe reservoir suggest momentum towards increased collaboration, our observations and conversations suggest that it is likely that management in Aiako Harria will continue in a monodominant paradigm where certain uses prevail over others unless there is considerable effort to reorganize arenas and address leadership. Maintaining the status quo appears to indicate a shift in use from a focus on timber management and traditional uses of land to one of biodiversity conservation. Evidence spanning from the variety of LIFE projects funded by the European Union to the Basque Government's efforts to increase conservation of parks and the creation the Euskadi Biodiversity Center, all suggest that upper levels of government and municipalities are taking a conservation-minded approach to management. However, the Gipuzkoa provincial government may still desire focusing on timber management, hunting, and fishing, in contrast to the changing focus of the other government entities.

If park governance does not attend to values feedback from various park user groups and create open arenas for inclusion and constructive debate, we envision that a paradigm shift to biodiversity conservation will continue to maintain entrenched perspectives and divisions between participants. Scientific management will dominate where a heavy reliance on biophysical data drives policy decisions. In contrast, an adaptive, holistic paradigm accounts for multiple methods of data collection, uses, and value outcomes.

ALTERNATIVES

To address the problem of integration in Aiako Harria, there are four options: the status quo, a top-down approach, a participatory approach, or a mixture of the two. In light of the analysis of trends and conditions above, the status quo will be dismissed as an option. Decision bottlenecks and alienation of user groups are not addressing the problem nor creating solutions.

A top-down approach would initiate at high levels of government likely at the European Union or at the Basque Government level. Generally, this approach does not account for the wide-ranging views of all affected stakeholders. Techniques in this style of management include information coordination, workshops for capacity building, job-training incentives, internship experience, and regulatory reform (Kahn, 1999). Through efforts to share information on best management practices, current scientific research, and progress towards goals, decisions can be made more transparent. Additionally, workshops are used as a forum for increasing knowledge and skills of government leaders in conducting integrated problem-solving. Finally, regulatory laws can be mandated and enforced through stringent economic sanctions, requiring decisions to be made in a particular way. In the United States, the National Environmental Policy Act (NEPA) requires that all government organizations designing policies must consider the impact to the environment. A similar law passed in the Basque Region could theoretically require that all projects be conducted consistently in compliance of particular criteria. However, pitfalls of a top down approach are numerous. Governing from a distance often results in a lack of consideration for local values and local conditions.

In contrast, a participatory approach to governance relies on the collaboration and inclusion of affected stakeholders in making decisions. This approach tends to be most successful when dealing with small-scale, local problems where trust and relationships can be built between individuals who otherwise may not cooperate. A technique called proto-typing is used where projects are initiated at a smaller size to reduce risk, to facilitate learning by doing, and to potentially maximize benefits. Successes from these projects can then be expanded and shared to provide a framework for larger scales in the future. In the participatory approach, workshops are also held between key players in the community and the directors of organizations to identify shared goals. Skill and knowledge sharing between diverse backgrounds and perspectives is emphasized, with a focus on building relationships and finding common ground within conflicting policies.

Issues with this approach include the challenge of incorporating diverse perspectives into integrated policies; focusing only on local issues, such as problems within park boundaries, while ignoring habitats not contained by political boundaries; and potentially missing funding opportunities from higher levels of government. To provide successful examples where top down and participatory approaches have been combined in effective multiple-use management, we will discuss two case studies: conservation and recreation measures in New Zealand, and the creation of the Boston Harbor Islands National Park in the United States.

Case Studies

In New Zealand, the Department of Conservation (DOC) manages recreation, parks, and conservation measures for the islands. The DOC desires to protect conservation values, to increase appreciation by people, and to reconnect people to conservation (Sutton and Cessford, 2007). In general, New Zealand government legislation guides park management practices. But regional and local management decisions incorporate community engagement through participatory crafting of regional and park specific conservation management plans (Sutton and Cessford, 2007). The DOC also actively involves community volunteers through a wide variety of activities including weed eradication, invasive species control, trail maintenance, facility upgrades, and coastal and river mitigation (Sutton and Cessford, 2007). These volunteer hours add up to an impressive 31,000 equivalent work-days; in fact, recent calculations suggest that over a 12 month period in 2003-2004, volunteer effort was equivalent to 63 full time staff positions (Sutton and Cessford, 2007). Not only do these volunteer opportunities save the DOC money, but also they build skills and knowledge of volunteers while connecting them to the land and to their community (Sutton and Cessford, 2007). Community initiatives and engagement are measured in DOC performance measures, recognizing that these actions increase *outcome ownerships* fostered in participants (Sutton and Cessford, 2007). However, managers acknowledge, "consultation is an opportunity for change, but change will only result when the engagement is meaningful and occurs at the right time and place" (Sutton and Cessford, 2007, p. 68).

In order to capitalize on meaningful engagement, New Zealand recently completed a Recreation Opportunities Review—the largest consultation process to date carried out on New Zealand's recreation facilities (Sutton and

Cessford, 2007). The purpose was to determine the existing range of recreation facilities and opportunities compared with what citizens desired: to decide which facilities required removal, needed improvements, or needed to be built to meet public need within the budget. The DOC first conducted a review and inventory of all existing facilities and then formed a national reference team made up of various outdoor recreation and tourist group representatives to determine how the process would occur, how decisions would be made, the principles the group would operate under, and how the public would be engaged (Sutton and Cessford, 2007). All of this was shared in a public document. Results from their detailed study included recommended management decisions for all facilities. The study was shared online and in published documents, soliciting public feedback on all proposed changes (Sutton and Cessford, 2007). Public meetings were held concurrently and thousands of feedback comments were submitted. After the entire process, an evaluation of people consulted was conducted, with 80% of respondents feeling that most or part of their recommendations were addressed in the final study conclusions (Sutton and Cessford, 2007). An important lesson learned from this process was that decision-making is not done through a formula, but rather, through negotiations around key principles of common interest (Sutton and Cessford, 2007).

Our second case study analyzes the creation of the Boston Harbor Islands National Park in the United States in 1996. Similar to Aiako Harria, an advisory council made up of representatives of various interest groups was created to help draft management plans for the park. To effectively engage the public, the park uses the following management techniques and methods: representation through partnerships and the Advisory Council, wide distribution of the Park's Draft General Management Plan, solicitation of public comments, partnerships with national/local environmental organizations, formal public workshops and meetings, legislated consultative meetings with Native Americans, outreach through community organizations, direct mailings using lists developed through community groups, educational outreach through the local school system, utilization of open partnership subcommittees, newsletters and the media, and informal networking with local minority groups (Makopondo, 2003).

To create the draft management plan for the Boston Harbor Islands Park, the National Park Service formed a planning committee to consult with the Advisory Council (Makopondo, 2003). They then held four workshops where over 400 people attended and hashed out issues and recommendations. This scoping phase resulted in the creation of two alternatives for the park: the alternative with the greatest overlap of common interest was selected, resulting in various management zones for different recreational activities (Makopondo, 2003). The conclusion of this management process was that involving diverse participants resulted in a consensus management plan that accommodated potentially conflicting issues (Makopondo, 2003).

Recommendations

In Aiako Harria, it is likely that a mixture of the top-down and participatory governance approaches will be necessary since success depends upon working within the constraints of existing organizational structures and culture. There are many policies and programs moving in the right direction that could be used as proto-type learning experiences. Concepts and ideas under development could be enhanced, encouraged, and expanded to create arenas for values and knowledge exchanges. For instance, efforts to use the Social Multi-Criteria Evaluation method to inform decision-making such as in the Urdaibai Biosphere reserve should be encouraged. The authors of that analysis recognized that top-down approaches were not sufficient in accounting for management challenges with high degrees of conflict where diverse scientists and stakeholders held differing interests and values (Garmendia et al, 2010). The application of this multi-criteria framework in a participatory context includes the following steps:

1. Identification of relevant social actors through institutional analysis, individual interviews with key agents and focus groups;
2. Problem definition;
3. Creation of alternatives and the definition of evaluation criteria. This process must be a collective process, resulting from dialogue between scientists and the social actors. Criteria are indicators that assess to what extent the different social actors' objectives are achieved by each alternative;
4. Valuation of criteria in a multi-criteria impact matrix. The matrix synthesizes the scores of all criteria for all alternatives. Each criterion score represents the performance of each alternative according to each criterion;
5. Assessment of social actors' preferences and values, mainly through in-depth interviews and focus groups;

6. Selection of the msulti-criteria evaluation method and application of the model through a mathematical aggregation procedure;
7. Social analysis, sensitivity analysis, and discussion of the results to check the robustness. (Munda, 2004)

Other examples of initiatives that should be encouraged include: community meetings, the concept of the Board of Trustees, the broad collaboration exhibited by the Añarbe Reservoir example, the draft public involvement plan, the information center built in Errenteria, the education and public involvement aspects of the Biodiversity Center, and the Irun trail restoration project. The municipality of Irun seems to be in a particularly good place to move forward as the council members have exhibited interest and support for conservation projects at the local level. Overall, the provincial government of Gipuzkoa may be in the best position to influence and facilitate this type of progress. From its oversight of the Park, the Board of Trustees is in a unique spot to be able to find common ground and to unite conservation issues at a larger scale with specific case-by-case issues on the ground.

Specific recommendations include:

1. Creating forums, workshops and public meetings for stakeholder feedback, focusing on opening lines of communication between user groups;
2. Actively engaging the public in participatory discussions of management plans for "sites of community importance" which include Aiako Harria
3. Using Agenda 21 in the municipalities to dialogue over common interest objectives between stakeholders
4. Increasing transparency of actions through explanations and notifications, including meaningful public feedback in this process;
5. Incorporating value rankings through social multi-criteria models into the decision process (Garmendia, Gamboa, Franco, Garmendia, Liria, and Olazabal, 2010);
6. Increasing the monitoring and appraisal of programs, creating a protocol for the termination of ineffective programs;
7. Measuring performance through benchmarks that include meaningful public participation;
8. Holding educational forums where community members can learn about biodiversity conservation, erosion control, invasive species, and

the cultural practice and significance of pollarding and charcoal making;
9. Soliciting feedback and information exchange through these forums to provide a two-way flow of information between decision makers and the community;
10. Re-instituting community work days where people come together to maintain trails, share picnics, create hunting blinds, learn traditional practices, and share knowledge;
11. Concentrating visitor-use on well-maintained trails or secure surfaces to avoid destroying habitat and to mitigate erosion (Cole, 1993);
12. Distributing park user surveys to assist in identifying specific outdoor facilities and forest management activities that attract or repel different types of tourists;
13. Using these surveys to identify and characterize all levels of conflict: offer answers ranging from "extremely desirable" to "extremely undesirable" to allow users to describe interactions with other groups in the park;
14. Understand goal interference by allowing users to answer "yes" or "no" to questions regarding the perceived interference of their recreational activities by the activities of other groups (Vaske et al, 1995)
15. If surveys suggest that goal interference is happening, zoning of use areas (management areas) is best for separating users (Vaske et al, 1995)
16. Using scientific surveys helps to clarify the standpoint of a greater number of people since those who attend meetings may not be representative of that user group (Vaske et al, 1995)

It is evident that a management shift to biodiversity conservation is gaining momentum and support in the Aiako Harria Natural Reserve. In addition to this dominant goal, the examples given above illustrate ideas to better integrate other values into the decision process, helping to better manage multiple goals. These efforts should always seek to fully orientate to and understand the problem, rather than assuming the problem is a scientific one. A multiple-use doctrine is not a recipe with a formula where you plug in recreation and public use and get out biodiversity conservation and sustainable harvesting. Rather, it is a process that includes soliciting the different ways people wish to use the land from the very beginning. These values can be incorporated into decisions via a framework like the Social Multi-Criteria

Evaluation model that was developed for the integrated coastal zone management of the Urdaibai Biosphere Reserve (Garmendia et al, 2010).

Furthermore, learning from the community-based activities of the past may enhance these methods. If people feel that the elements of the park they value are being provided for, they are more likely to "buy-in" and actively work to maintain and preserve them. Reintroducing community workdays could bring groups together who may otherwise not agree. Emphasis on cultivating and exchanging values of respect, skill, and affection between individuals would go a long way toward managing in the common interest.

Concurrently, the management of Aiako Harria requires the unification of park goals such that policies are clarified and made consistent across ownership boundaries. With five municipalities, each with their own separate rangers and agenda, it is easy to understand the prevailing confusion around park management. Such actions will help foster a sense of ownership and buy-in by decision-makers, employees, and users of the park.

Conclusion

In Aiako Harria, governance is not adaptive and integrated, resulting in management practices that are not in the common interest. Fragmented ownership, multiple layers of governance, and conflicting goals and objectives are serving few people well. Many innovative ideas from above and below stagnate at the provincial government level due to value deprivations of wealth and power and contradictory goals. Furthermore, conflict between park user groups at the ground level adds to the instability and distrust. Action needs to be taken to build on efforts to manage for the common interest, shifting to a multiple-use management doctrine.

References

BAP (2009). Europa, Gateway of the European Union. EU Biodiversity Action Plan. Retrieved November, 2009 from http://ec.europa.eu/environment/nature/biodiversity/comm2006/index_en.htm.

Brewer, G. D. (1983). The policy process as a perspective for understanding. In E. Zigler, S. C. Kagan and E. Klugmand (Eds.), *Children, families and government*, Cambridge: Cambridge University Press.

Brunner, R. D., T. A. Steelman, K. Coe-Juell, C. M. Cromley, C. M. Edwards, and D. W. Tucker. (2005). *Adaptive Governance: Integrating Science, Policy and Decision Making*. New York: Columbia University Press.

Clark, S. G., Cherney, D. N., Angulo, I., Bernardi De Leon, R., Moran-Cahusac, C. et al. (2009). A Problem-Oriented Overview of Management Policy for Podocarpus National Park, Ecuador, *Journal of Sustainable Forestry*, 28:6,663-679.

Clark, T. W. (2002). The Policy Process: A Practical Guide for Natural Resource Professionals. New Haven: Yale University Press.

Cole, D. (1993). Minimizing Conflict between Recreation and Nature Conservation. Ecology of Greenways: Design and Function of Linear Conservation Areas. Minneapolis: University of Minnesota Press. Council Directive 92/43/EEC of 5/21/1992 on the conservaton of natural habitats and of wild fauna and flora. Habitats Directive.

Cromley. C. M. (2002). Beyond Boundaries: Learning from Bison Management in Greater Yellowstone. *Doctoral dissertation, Dissertation.* Yale University, New Haven, CT.

Errenteria Town Council Website. Retrieved November, 2009 from http://www.errenteria.net/es/html/9/429.shtml.

Euskadi Biodiversity Centre, (2009). Retrieved October, 2009 from http://www.torremadariaga.net/.

Garmendia, E., Gamboa, G., Franco, J., Garmendia, J. M., Liria, P., Olazabal, M., 2010. Social multi-criteria evaluation as a decision support tool for integrated coastal zone management. *Ocean and Coastal Management* 53, 385-403.

Gizpukoa Provincial Government, (2009). Retrieved December, 2009 from http://www.gipuzkoa.net/.

Irun Town Council, (2009). Retrieved November, 2009 from http://www.irun.org/cod/inicio_especifico.aspx?tipo=3andidioma=1.

Kahn, H. (1998). The Prospects for Integrated Watershed Management in the Panama Canal Watershed. *Journal of Sustainable Forestry*. 165-180.

Makopondo, R. (2003). Collaborative recreation and tourism planning within a national park context: the process, stakeholder identification, and expected outcomes. Proceedings of the 2002 Northeastern Recreation Research Symposium. *Gen. TeChapter Rep*. NE-302. Department of Agriculture,

Forest Service, Northeastern Research Station. Newtown Square, PA: U.S. 201-207.

Munda, G. (2004). Social multi-criteria evaluation: Methodological foundations and operational consequences. *European Journal of Operational Research* 158: 662-677.

Lasswell, H. D. (1970). The emerging conception of the policy sciences. *Policy Sciences* 1: 3-14.

Lasswell, H. D. (1971). A Preview of the Policy Sciences. New York: American Elsevier Publishing Company.

Lasswell, H. D. and S. McDougal. (1992). *Jurisprudence for a Free Society: Studies in Law, Science, and Policy*. New Haven: New Haven Press.

Life Aiako Harria (2009). Retrieved November, 2009 from http://www.lifeaiakoharria.net/.

Perez, J. D. and Groome, H. (2000). Spanish forestry planning dilemmas: technocracy and participation. *Journal of Rural Studies*. 485-496.

Regional Council Decree 46 of May 4, 1999 on Aiako Harria Natural Park Management. Regional Council of Gipuzkoa. Retrieved January, 2010 from http://www.ingeba.org/lurralde/lurranet/lur24/erlaitz/df46.html.

Sutton, S. and Cessford, G. (2007). Community involvement in planning and management for outdoor recreation in New Zealand protected areas. Science and stewardship to protect and sustain wilderness values: Eighth World Wilderness Congress symposium; September 30-October 6, 2005; Anchorage, AK. Proceedings RMRS-P 49. Fort Collins, CO: U.S. Department of Agriculture, Forest Service, Rocky Mountain Research Station. p. 65-73.

Vaske, J. J., Wittmann, K., Laidlaw, S., Donnelly, M. P. (1995). Recreation conflicts on Mt. Evans. Proceedings of the 1994 Northeastern Recreation Research Symposium; 1994 April 10-12; Saratoga Springs, NY.: *Gen. TeChapter Rep*. NE-198. Department of Agriculture, Forest Service, Northeastern Forest Experiment Station Radnor, PA: U.S. 96-99.

Walker, B., Holling, C. S., Carpenter, S. R., and Kinzig, A. 2004. Resilience, adaptability and transformability in social-ecological systems. *Ecology and Society* 9(2):5.

Watson, A. (2001). Goal interference and social value differences: understanding wilderness conflicts and implications for managing social density. Visitor use density and wilderness experience: proceedings; 2000 June 1–3; Missoula, MT. Proc. RMRS-P-20. Department of Agriculture,

Forest Service, Rocky Mountain Research Station Ogden, UT: U.S.: 62-67.

In: Multiple-Use Management … ISBN 978-1-61942-328-2
Eds: A. DuVal, T. Muruetagoiena et al.

Chapter 2

Mountains to Sea: Riparian Rehabilitation as a Multiple Management Strategy in the Basque Region, Spain

Michael J. Blazewic [*]
Environmental Management at the Yale School of Forestry and Environmental Studies, New Haven, CT, USA

Abstract

A programmatic focus is offered for the rehabilitation of riparian areas in the Basque Country of Northern Spain. Restoration of functioning riparian buffers would improve aquatic ecosystems, water quality, and floodplain function in order to address the European Union Directives in Water, Habitat, and Flooding. This chapter explores the possible benefits of such an effort through the example of the Oiartzun River. Management strategies focusing on the ecosystem services offered by riparian forests, floodplain wetlands, and headwaters forests (particularly those located in Aiako Harria Park) are reviewed. Case

---

[*] Email: michael@roundriverdesign.com. The author would like to thank our Basque Country hosts for organizing an informative and enjoyable visit. Thank you to Ibon Tamayo who provided assistance with procuring GIS data. The author would also like to thank ProfessorMark Ashton for his guidance in developing the content of this paper in addition to the author's classmates for their support and review of his drafts.

studies of multifunctional landscape development for riparian and floodplain management are offered. These examples highlight green infrastructure solutions that have successfully provided clean water, biodiversity, flood reduction, green space, recreation, and other benefits to watershed communities as productive alternatives to intensive channel management and as an important progression for post-industrial urban redevelopment.

Keywords: Aiako Harria, floodplain, wetland, geomorphology, buffer, flood reduction, watershed management, Oiartzun River, water resources, aquatic biodiversity, green infrastructure

INTRODUCTION

Aiako Harria Nature Park is part of the European Union Natura 2000 network of designated areas of conservation and protection. The Park rests at the headwaters of the Urumea, Oiartzun, and Bidasoa rivers in the Basque Country of Northern Spain. These rivers descend rapidly from Park forests and pastures transitioning into broader residential and agricultural valleys, eventually becoming highly urbanized and degraded as they reach the Bay of Biscay. The ecological health, water quality, and flood intensity of these streams are heavily dependent on the quality and quantity of riparian areas that act to buffer upland impacts. These transitional lands are among the most diverse, dynamic, and complex habitats on the terrestrial portion of the landscape (Naiman and Décamps, 1997). Riparian areas are, however, notoriously difficult to define and delineate due to their complex three-dimensional boundaries. That is, the boundaries extend down into the groundwater, up into the canopy, outward across the floodplain, up the slopes that drain to the water, and parallel along the stream at varying width (Verry *et al.*, 1999).

In the Atlantic region of the Basque Country the presence and function of riparian forests and wetlands have been historically undervalued as evidenced by extensive clearing, draining, and filling of these environments, which has left them exceedingly rare. It is evident that a multitude of resources (financial, material, time, and energy) have been spent protecting property adjacent to rivers by methods such as channel straightening, dredging, and streambank armoring. These practices often are short-lived, degrade aquatic ecosystems and water quality, and may even increase flood hazard dangers in downstream reaches.

A holistic approach to improve stream condition recognizes lotic ecosystems as needing continuity, both laterally across floodplains and longitudinally from headwaters downstream to their mouth (Schumm, 1977; FISRWG, 1997). It also recognizes that even the most pristine streams are inherently dynamic across space and time (Rosgen, 1996; Montgomery and Buffington, 1997) as are their riparian forests (Verry et al., 1999, p. 238). Building upon these themes, and out of request by the Euskadi Biodiversity Centre to investigate how to improve forest management in relation to stream ecosystem health, this paper investigates current conditions and recommended land management practices at three major zones within the watershed: headwaters, transition, and deposition. It is argued that reconnecting the lateral and longitudinal riparian areas through these zones would reduce the need for on-going channel management, improve ecology and water quality, and shift the focus of management projects from short term control (5-10 year planning) to long-term equilibrium and stability (50 to 150 year planning).

As watershed science has evolved so also has policy. Recent policy mandates are led by EU Directives including the Water Framework, Habitats, and Floods Directives, which have asked the Basque government to improve management of rivers. Individually these laws are complex and multi-faceted. It is the further purpose of this paper to recognize the importance of policy and technological tools and offer strategies for accomplishing improved water quality through off-site methods. It is proposed that the rehabilitation and management of riparian areas and floodplain wetlands would provide a programmatic focal point with which to address components of all three of these Directives in an immediate and meaningful way. Doing so, I argue, would also provide valuable ecosystem services and secondary benefits for the human community above and beyond legal requirements.

In order to illustrate these ideas further the Oiartzun watershed is presented as a case study. The Oiartzun River was chosen for several reasons. The Park boundary encircles a significant portion of the total Oiartzun watershed area (Figure 1) and provides a greater percentage of drainage area to the Oiartzun than either the Urumea or Bidasoa Rivers. For this reason the Park provides a significant influence on the quality of the Oiartzun River and sets a precedent for land management that ultimately will affect those municipalities which lie downstream: Oiartzun, Errenteria, Lezo, and Pasaia. Also, of the three primary watersheds originating in Aiako Harria Park, the Oiartzun is the most degraded and perhaps most in need of attention and repair.

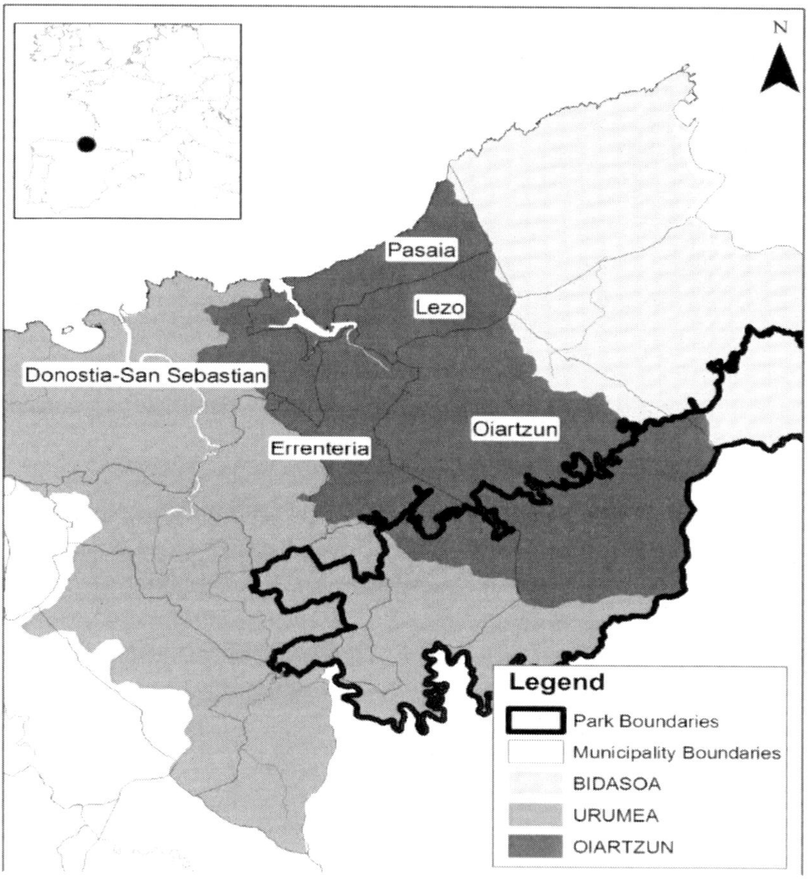

Figure 1. Project location map: major watershed and municipal boundaries of Gipuzkoa province in relation to the Aiako Park and municipalities.

During our field visit to the Basque Country we formally met with twenty-one professionals. These individuals work for private and public organizations, public companies, and local and provincial government agencies. In addition to our guided tour and interviews, I paid extra attention to riparian areas and conducted rapid visual surveys of stream conditions while driving and walking in the region. Further development of this paper was conducted remotely by utilizing primary and secondary literature, in-country contacts, and GIS-aided computer analysis.

WATERSHED DESCRIPTION

Located in the western Pyrenees Mountains near the border between Spain and France, the Oiartzun watershed escaped glaciation events that scoured lands in the higher peaks to the east. Instead regional surficial geology has undergone significant weathering associated with abundant precipitation and extensive human land clearing (Remondo et al., 2005). Precipitation in the watershed is frequent (especially in the winter months) due to maritime and topographic influences. This area receives the highest rains of the Basque Country: the annual precipitation surpasses 2,000mm/year in valleys, and at the highest peak of Aiako Harria (832 meters elevation) rainfall can exceed 2,800 mm/year. Draining from the high plutonic dome that constitutes the bedrock core of the Park center, water moves downhill rapidly connecting into streams with very high hydraulic energy values (Docampo et al., 1991). Especially in the headwaters of the Park high bed-load streams have extensively down-cut V-shaped valleys.

Stream gauging stations indicate that regional runoff coefficients are highest in the Urumea, Bidasoa, and Oiartzun River watersheds and that these streams produce the greatest volumes of runoff per hectare in the Basque Country (P. Departmento de Medio Ambiente, Planificacion Territorial, Agricultura y Pesca, 2006). This runoff is a result of the high rainfall as well as extensive human alteration of the landscape. Lack of water-bearing aquifers in the region and steep slopes further reduce natural water retention potential of the landscape. Records indicate flooding occurs more frequently during the winter months when evapotranspiration is reduced and precipitation is typically greater. Periodic development of a winter snowpack may also heighten flood potential. Despite relative abundances, hydrological droughts are recorded on average 21% of the time (Basque Water Agency, 2006). This shifting drought/flood cycle is indicative of both lack of landscape holding capacity as well as seasonal variability in precipitation.

In total the Oiartzun watershed covers 85.6 square kilometers and has a main channel length of 19.8 km. The Sarobe (8.5 km, 18.18 sq. km) and Karrika (9 km, 9.4 sq. km) are the main tributaries. Near the settlement of Ergoien, the valley broadens and flattens allowing a meandering channel and more substantial floodplain to develop. The river in this area and upstream has been reported as having "good ecological condition" based on physical, chemical, and qualitative data regarding vegetation, land use, and appearance (Ramirez, 2005; Borja et al., 2003). Downstream where the Karrika tributary enters in Ugaldetxo, the river becomes almost completely channelized and nearly all traces of the floodplain and natural meander bends have been erased. This lower portion of the Oiartzun is ecologically impoverished, the once thriving estuary having been transformed into the almost completely industrialized Port of Pasaia (Ramirez, 2005; Borja et al., 2003). So extensive is the degradation near the estuary that a 2008 report issued by Greenpeace International cited serious contamination issues due to extensive urban and industrial pollution ("Destruction at all Co[a]st: 2008: Report on the Spanish Coast Situation" n.d.). Another analysis of pressures in the Atlantic coastal estuaries concluded that the Oiartzun River is at high risk for failing to achieve the condition of "good ecological status" within the Water Framework Directive objectives by 2015 due to development and industrial affects on water and sediment pollution, shoreline manipulation, and intertidal wetland loss (Borja et al., 2006).

CONTEXTUAL CONSIDERATION FOR WATERSHED MANAGEMENT

In a simplified model watersheds are described as having three parts, the headwaters or source zone, mid or transfer zone, and the lower or depositional zone (see Figure 2 based on Schumm, 1977). These zones provide an idealized structure to understand key watershed processes such as sediment transport, hydrology, channel morphology, and ecology (the "river continuum" concept developed by Vannote et al. (1980) describes this in greater detail). While particular systems may deviate, in general the Oiartzun River follows this model closely as described in the previous section of this paper: steep headwaters flow to a gradually widening flatter valley eventually leading to a depositional zone at the estuary (Figure 3).

As recently reported by The Nature Conservancy, understanding the components and connectivity of each watershed component is critical for managing for river health (Smith et al., 2008). Land use activities in Aiako Harria Park (and the example they demonstrate for other landowners) have significant impacts on water quality and quantity. Management along the entire continuum is necessary as each element supports the other. The upper watershed area for instance plays a critical role in contributing inorganic and organic material, including sediment, coarse particulate organic matter, and large woody debris to the river system. In fact, small headwater catchment processes and inputs support biological diversity all the way downstream to the mouth of a river (Meyer et al., 2007).

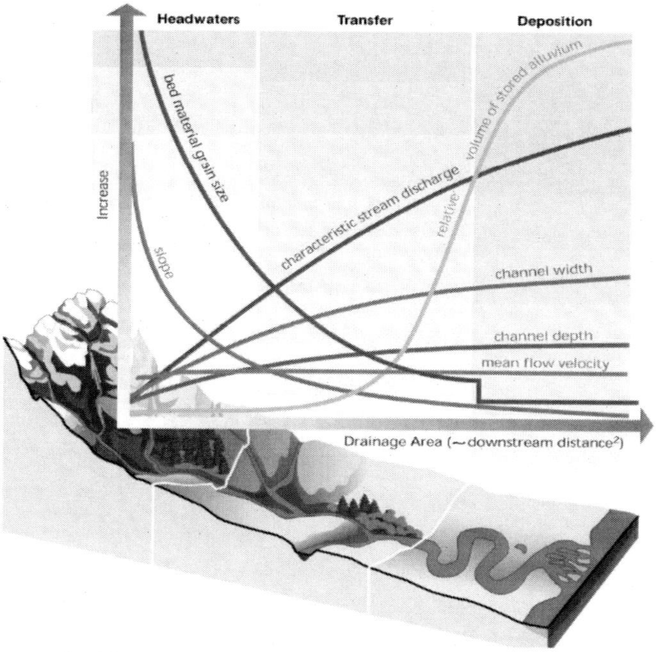

Source: Stream Corridor Restoration: Principles, Processes, and Practices, 10/98, by the Federal Interagency Stream Restoration Working Group (FISRWG).

Figure 2. Simplified model of longitudinal zones of a river and their generalized characteristics.

In addition to longitudinal connectivity, healthy streams maintain important connections perpendicular to the direction of flow. Such is the case when geology allows (i.e. not bedrock or clay) development of a hyporheic zone (region of mixing shallow groundwater and surface water) to extend down and out into the sediments surrounding the stream channel. This area is important to aquatic biota, fish spawning, and water chemistry and is minimized where streams are channelized and armored. Channel connectivity to adjacent riparian areas and floodplains is also critical for aquatic ecosystem health and channel stability. In contrast, management activities focusing on the wetted channel fail to account for the temporal nature of streams and the critical interplay that occurs between the aquatic and terrestrial ecosystems within the riparian and hyporheic zones.

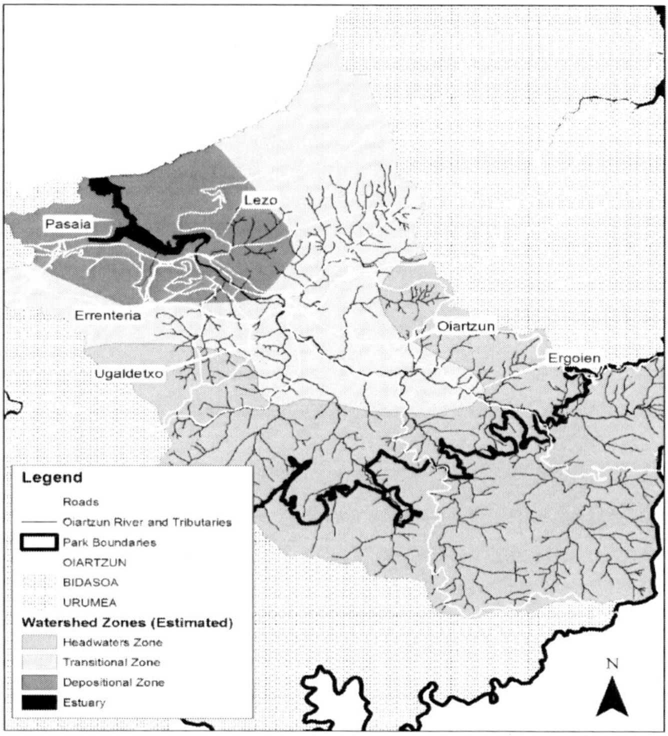

Figure 3. Generalized map of Oiartzun watershed transition zones along with major roads and settlements.

Where riparian vegetation is removed and woody debris not recruited, stream depth and velocities increase and the hyporheic zone narrows. Reduced subsurface flow through the channel banks and bed reduces nutrient removal and may flush excess particulate matter into the system. Several studies from the region have investigated the multitude of benefits gained by having large woody debris in streams (Elosegi et al., 2008; Diez et al., 2001) due to their role in slowing and spreading water as it flows down a stream channel.

Also relevant to the management of Aiako Harria is that while the larger lowland stream channels gather and deserve attention (because they often have been the most impacted), in terms of contact with water and water quality improvement, priority should be given to maintaining and installing buffers on smaller streams (first and second order) rather than on higher order streams (see Strahler, 1952 for a description of stream orders). This is because the total length and percentage of cumulative length of ephemeral, first, and second order streams is magnitudes higher than along large high order streams. If riparian forests are only established on high order (lowland, wide) streams then a great deal of opportunity to intercept and interact with the water in the channel has already been lost (National Research Council, 2002).

Restoration Challenges and Opportunities

Headwaters Zone

While human land use and alteration can be traced back centuries in the Basque Country (see Cousins and DuVal this volume), managers today are predominately faced with addressing more recent legacies. One legacy, the significant deforestation that occurred by the start of the 19th century, has especially affected the region. Following this clearing, planting of exotic conifers was encouraged to prevent soil erosion and develop pulp and timber markets in the mid-1900's (see Caligiuri this volume). The impact of cutting, especially on steep headwaters slopes and in riparian areas remains significant. On public and private land, forest practices such as scalping and down-slope ripping to prepare for *Pinus radiata* (D. Don) planting is known to cause significant erosion and increase runoff (Olarieta et al., 1999) as is windrowing slash perpendicular to the slope (a common practice our group observed). After harvesting timber, water runoff from the site can increase due to less leaf interception and reduced evapotranspiration leading to increased peak flows downstream (Langer et al., 2008). Furthermore, exposed soils are susceptible to compaction and greater disturbance thus reducing their overall quality. While timber harvesting for economic gain has recently slowed due to a recession in the market, cutting in Aiako Harria to transition forest back to native species is on-going and may be a threat to water quality and quantity if done inappropriately (e.g., clearcutting steep slopes and riparian areas).

Forestry activities are not necessarily detrimental altogether, however. Managing upland forests for water quality, for example, may necessitate active silviculture prescriptions. Uneven-aged and mixed species stands maximize nutrient sequestration and minimize risk to water quality that would result from large scale disturbance such as pests, fire, or windthrow (Barten et al., 1998) (See Table 1 for a further description of contributions from forests to streams during successional stages). Harvesting near or within riparian areas is also possible so long as exceptional planning and harvesting techniques (including use of special equipment) are carried out.

Table 1. Summary of input contributions from forests to streams during various forest successional stages. Adapted from Verry et al. (1999)

	Successional Stage				
Contributions from forests	Initiation (Yr 0-10)	Organization (Yr 10-30)	Aggradation (Yr 30-80)	Mature (Yr 80-125)	Old Growth (Yr 125-250+)
Water yields and peak flows	Increased	Stable			Decreased
Erosion and sedimentation	Increased	Variable			Decreased
Nutrient leaching	Increased	Minimal			Gradual Increase
Water Temperature	Increased	Stable			Decreased
Organic Matter	Greatly Reduced	Gradual Increase, Changing Quality			Steady State

A1)

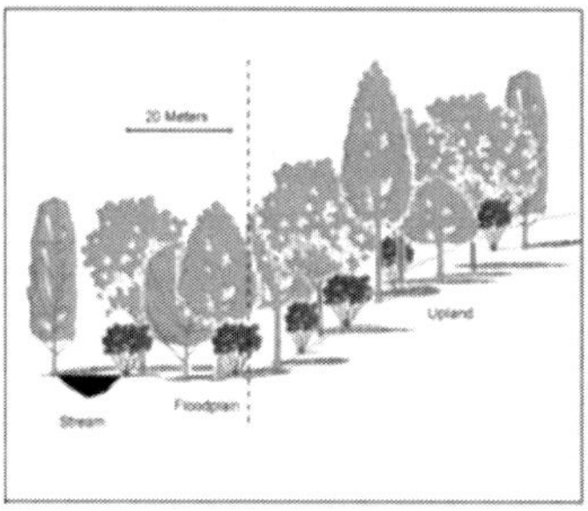

A2)

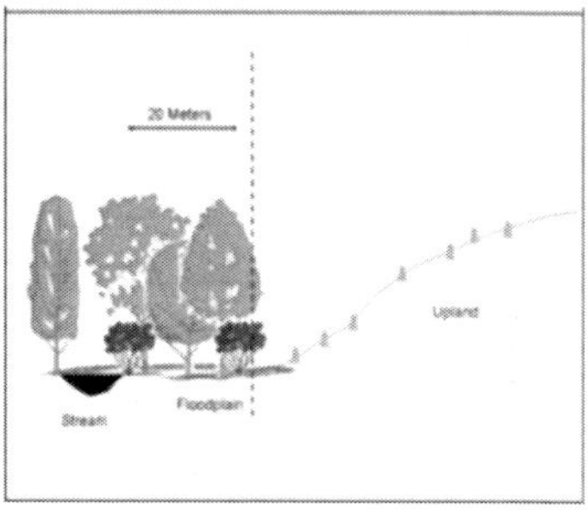

B1)

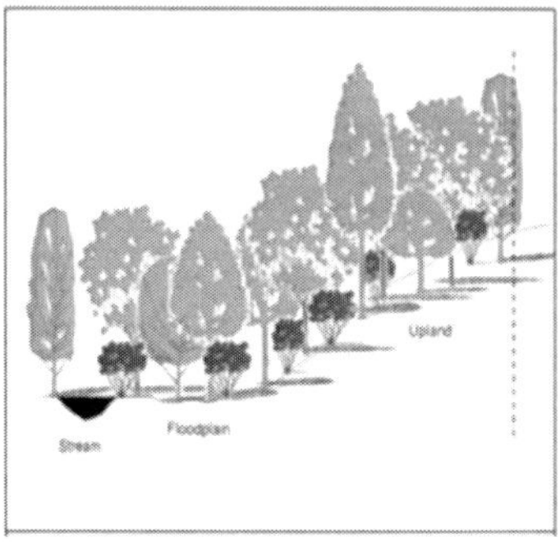

B2)

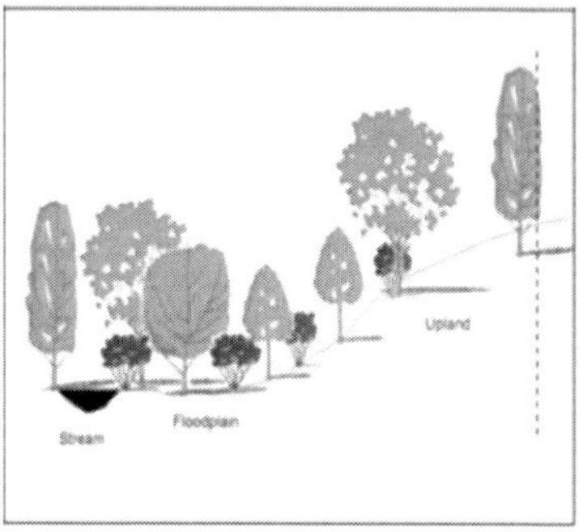

Figure 4. Graphic contrasting a strict riparian management boundary and the potential for harmful silvicultural activities within close vicinity of the riparian area (A1 and A2) compared with a riparian management boundary line at the upland of the ecotone and a gradient of silvicultural and best management practices within the vicinity of the riparian area (B1 and B2). Developed based on graphic in Verry et al., 1999.

Verry et al. (1999) recommend adhering to three ecological principles for riparian silviculture. First, riparian forests vary significantly in structure and composition and thus require equally diverse harvest approaches. Second, riparian function is highly dependent on structure and species composition. Third, they are transitional areas and thus not easily delineated.

This last principle is often ignored where riparian setbacks are enacted (twenty meters in Aiako Harria Park) leaving the area inside the boundary uncut (low impact) while the adjacent area may be clearcut to regenerate an autochthonous naturalized or planted exotic stand (high impact). Often the boundary may not adequately protect riparian functions (for instance where intermittent tributaries or seeps and wetlands connect) but instead lead to sharp boundaries with significant edge effect between upland areas and what are the perceived, but not always correctly delineated, riparian zones. A more appropriate approach would delineate a riparian management zone using field-surveyed information and allow for a gradient of silvicultural activities that maintain riparian function while also allowing for timber harvest (see Figure 4). Also, since within headwaters streams large woody debris recruitment is beneficial, wood may be either added through well planned felling towards streams, or ideally, through natural recruitment (Elosegi et al., 2008).

Away from the riparian zone land management embraces multiple objectives (Figure 5). Forest management shifts to maintenance of uneven aged stands via patch selections and irregular shelterwood systems. Such forestry would work to promote native species diversity through appropriate guilds (e.g., mixed-beech and mixed-oak forests as described by Caligiuri this volume) as well as promote resiliency through diversity.

Further away pollard groves, orchards, and other agroforestry techniques are well suited as the maintenance of vigorous growth and of year round roots promotes soil stability and may even improve soils (Gordon and Newman, 1997). Further away from the riparian areas still, along drier ridgelines and flatter slopes more intensive management activities can be pursued with less risk to water quality. Here open fields for grazing, hunting, recreation, and/or aesthetics may be maintained as well as more intensive silvicultural practices.

A secondary affect of timber harvesting is road building, which appeared pervasive in the Park. A study of roads in a Douglas Fir (*Pseudotsuga menziesii* L.) plantation in the western United States describes peak flow increases due to improved flow pathways created by roads. Debris flows from erosion, which was evidenced during our site visits in the Basque Country, were also found to be directly linked with road networks (especially across hillsides) in this Oregon experimental forest (Jones et al., 2000). Road culverts, especially hanging culverts that are formed across steep slopes, are notorious for creating geomorphic instability (Riley, 1998, p. 358) as well as creating barriers to movement for aquatic biota. Furthermore, roads have been linked with assisting invasive species dispersal (Parendes and Jones, 2000). Inventory and decommission of non-essential roads and removal and restoration of culverts should be part of the overall habitat restoration program in Aiako Harria. This is especially true where vegetated wetlands may appear at the transition between convex and concave slopes (i.e. the keypoint), an area where roads are often placed. Wetland reconstruction efforts should seek to develop diversity in edge, depth, cover, and material (i.e. rocks, stumps, wood) in these wetlands in order to promote greater biodiversity.

Grazing is widely practiced in headwaters streams and is another agent of disturbance of riparian forests and understory vegetation. Riparian vegetation stands little chance of developing age class and species diversity when grazing pressure exists (National Research Council, 2002). While removal of livestock from all riparian areas and/or fencing would be ideal, this solution is impractical due to the abundance of streams in the Park.

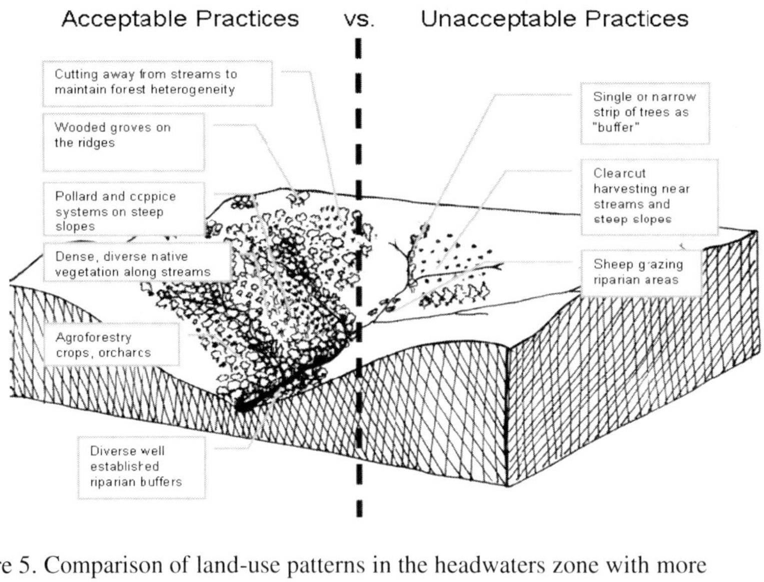

Figure 5. Comparison of land-use patterns in the headwaters zone with more acceptable practices stylized on the left as compared with those practices that lead to riparian and water quality degradation on the right.

Instead a plan should be developed to prioritize the most critical riparian areas. For instance while on a hike in the Park our group noticed horses grazing down to stubble riparian vegetation and standing in a stream that was posted as a special trout fishery. Fencing and providing alternative watering sources along these most important riparian areas would allow natural regeneration and the development of functional density and diversity.

Another management alternative would be to implement stricter regulations or develop a permit system that would require shepherds to monitor animals and keep them from grazing riparian zones. The benefits of such management will extend beyond water quality. For instance Guiliano and Homyack (2004) found small mammal populations increased as a result of fencing livestock. A last suggestion is that private riparian lands be swapped for public land on drier ridgelines. Putting riparian land into public control where it can be managed as a community forest for watershed protection and biodiversity would work to achieve improvements for downstream watershed residents while at the same time allowing for greater likelihood of a private landowner being able to economically manage drier, more arable ridge lands.

In summary, a combination of management techniques is needed in order to yield a pattern of land use for the headwaters area (predominately the land of Aiako Harria Park) that acknowledges and incorporates a working landscape while shifting to a new paradigm of biodiversity enhancement and water quality and quantity provision.

Transitional Zone

As is typical in a densely populated watershed (2150 inhabitants per km^2) historic alteration of stream channels has had a great impact on the Oiartzun River. Impacts from channel straightening and riparian forest removal affect nearly all of the reaches downstream of Ergoien (Ramirez, 2005). Channel straightening increases the slope and therefore the power of a stream. This increase in stream power is typically followed by channel incision and eventually widening. Widening is often prevented by bank armoring with stone or concrete, however, effectively “locking” an incised channel in place and preventing floodplain access (Brierley and Fryirs, 2008; Sweeney, 2004). Additionally, development encroaching onto the floodplain of the Oiartzun River (e.g., buildings, roads, paths) effectively raises the bank height which has the effect of raising flood water levels in surrounding areas. Rivers that do not have the pressure relief of a floodplain have increased channel power and are at risk of catastrophic fluvial erosion episodes.

Channel straightening has an impact on sediment transport as well. During high flows, small sediments are easily transported and deposited where low velocities are found (typically on the inside of a bend or the floodplain). When floodplains do not exist or are inaccessible and where bends have been removed through straightening (as has occurred along much of the lower Oiartzun) fine sediments may be transported long distances (in this case all the way to the estuary at the mouth of the river). As fine materials have the highest concentration of nutrients and organic material, the absence or overabundance of fine sediment can have great impacts on the aquatic biology of the stream and the receiving body (which in this case already suffers from high nutrient inputs). Similarly, large sediments such as cobbles and gravels, which set up habitat features and provide long-term channel stability and erosion resistance, may be displaced leading to fluvial erosion hazards.

From a habitat perspective, channel straightening affects channel bed heterogeneity and thus aquatic organism populations. For the *Galemys pyrenaicus*, the desman, cold oxygen-rich water and abundant aquatic insect populations are a requirement (Palmeirim and Hoffmann, 1983). A 2006 study in Biscay found that for the European Mink, a species in significant decline across Europe, two main factors determined its absence, "polluted waters and canalized streams" (Zabala et al., 2006).

Another significant impact to the hydrologic regime of the Oiartzun River is the alteration of landscapes. Clearing of forests and the draining of fields and wetlands can decrease water storage capacity and cause an increase in surface water runoff timing and peak flows. This runoff is typically high in non-point source pollutants which are problematic in the Oiartzun (Borja et al., 2003). Stormwater inputs from roads, drainage ditches, and impervious surfaces (such as parking lots, rooftops, sidewalks, and paving stones that dominate the village centers) quickly convey water into the river. These alterations increase the timing and amount of water (both peak flow and total volume) entering the channel during a runoff event and may contribute to localized channel incision, enlargement, and/or downstream flooding.

In order to reverse the trend of increased runoff, rehabilitation of and investment in green infrastructure should be considered in the transitional zone of the watershed. Green infrastructure refers to the living systems of a landscape, such as riparian forests, that provide valuable community services. Design considerations for riparian zone rehabilitation are numerous and no one form, width, or species palette fits all. Different buffer widths may achieve different objectives. For instance a width of 25 to 55 feet might achieve bank stabilization objectives while a width of 100 to 300 feet may be necessary to maintain habitat for songbirds and large mammals (Klapproth and Johnson, 2001). For streambank stability, temperature control, and pollutant removal, substantial benefits are achieved within the first 50 feet of vegetated buffer (Rupprecht et al., 2009). A better approach than designating a strict width is to establish the objectives for the buffer program and to base buffer widths on available technical, scientific, economic, and social requirements of the project through committed leadership and involvement of key stakeholders (Verry et al., 1999). Perhaps even more important than buffer width in regards to water quality is buffer continuity parallel to a stream. According to Weller et al. (1998), gaps, including road crossings, in riparian buffers provide access points through which nutrients from upland areas can be easily discharged into streams.

In transitional zones (as well as headwaters and depositional zones) if the goal is to reduce peak flood flows then the vegetation type and distribution of riparian vegetation is very important. The main contribution to reduction of flood velocity and quantity is vegetation roughness. Shrubs that flatten over easily during high flows do not provide significant friction to a fast moving stream. Thomas and Nisbet (2007) note, "the spacing and layout of trees, smoothness of trunks, presence of lower branches, level of undergrowth and amount of dead wood on the woodland floor all have an effect," and thus design, installation, and management of riparian forests for increased roughness would contribute to their capacity to slow and store flood flows.

Along farm clearings, or even healthy riparian forests, a stepped approach using short rotation coppices of willow (*Salix sp.*), hazel (*Corylus avellana* L.) and black alder (*Alnus glutinosa* L. Gaertn) may provide some additional economic diversity to farms while also softening the edge between low growing crops and the river. A paper examining their potential in Sweden recommends managing coppices along farm fields for their ability to attract birds and pest-reducing insects, their ability to increase local biodiversity, and to "catch" pesticides and nutrients that may leech from fields (Berg, 2002).

Wetland appearances may differ significantly (i.e., open water, forested, hydro-connectivity) across river systems yet these features may be capable of providing many of the same benefits and therefore should be considered intragral components of riparian forests. Benefits may include wildlife and fisheries habitat, sediment and nutrient retention and export, flood control, groundwater replenishment (helping to maintain baseflow to rivers), water purification, heightened biodiversity, non-timber forest products, recreation and tourism, and landscape aesthetic. In mid-elevation reaches, floodplain wetland rehabilitation should focus on opportunities to redevelop altered sites (using historic maps and aerial photographs). Another popular and relatively inexpensive technique is the plugging of field drainage ditches in order to promote the capture and storage of water on the floodplain. This may develop ephemeral or permanent wetlands depending on soil structure and height of the water table. Ecosystem service payments to landowners for allowing development of these features have been successfully demonstrated in the northeastern United States.

A possible model for urban stream renewal is a LIFE-funded project of a highly urbanized and altered stream system, the River Liesing in Vienna, Austria. Rehabilitation activities included construction measures to restore river continuity, the restoration of semi-natural morphological conditions and former meanders, the construction of a semi-natural river bed with a gravel substrate, and the restoration of the river's natural transport capacity. A pathway and a riverside child's playground were also constructed. The project has resulted in the creation of aquatic ecological corridor through the urban area with improved habitat for priority species along with benefits to the human community ("EUROPA - Life Projects - Project Summary" n.d.).

Depositional Zone

Like many coastal river mouths the Oiartzun's estuary has become heavily industrialized. A 2005 report indicated 24,164 companies (including industry, energy, construction, commerce, transport, and services) operating near the estuary of the Oiartzun River. That same report indicated that 66.8% of the shoreline had been reinforced with steel and concrete walls (Borja et al., 2006). Ship building has been intensive due to the unique natural features that create an excellent port. Paper mills, metal fabrication and processing, and textiles manufacturing are also found in the valley (Borja et al., 2003). The Arditurri mines are another industrial legacy that have had a negative effect on the river due to leaching from tailings piles which has transported heavy metals downstream with alluvial sediments. A 1995 study found that these pollutants had been deposited downstream on some of the areas agricultural fields (Ansorena et al., 1995). Here particularly, a filter of riparian vegetation may have trapped these pollutants from both leaving the site as well as from being washed onto the farm fields.

Despite this legacy of pollution and alteration, in recent years there are signs pointing to improvement in the health of the local environment. In 1996, new municipal upgrades to treat sewage and industrial pollution led to a significant enhancement in water quality of the estuary (Borja et al., 2009; Uriarte and Borja, 2009). Within this same area a transition to a post-industrial phase for the municipalities is expected to take place (Borja et al., 2006). This shift away from an economy that relied heavily on industry may provide opportunities to improve riparian areas and flood control through brownfield renewal projects that address new community priorities such as open space, quality of life, environmental health, and recreation. One such project that combines urban renewal with estuary improvements is found nearby at the Bidasoa River mouth at Txingudi. The creation of wetland features integrated into an attractive park for residents appears to be an ideal project for providing critical coastal habitat and providing water purification to the estuary. A similar project would be a terrific starting point for the restoration of the Oiartzun.

A LIFE-funded project on the Guadiato River (Spain) may provide some additional guidance on working to restore and integrate water resources in a post-mining industrial area. In the Guadiato, project partners are using phytoremediation to prevent erosion and contamination, as well as restore the landscape of an old mine's dumpsite. The project has also strived to stimulate stakeholder interest in the river's ecosystem and in the benefits of environmentally responsible behavior, such as building waste-treatment plants, and drafting and implementing environmental action plans ("LIFE Guadiato Report" n.d.). At the center of these shifting priorities, stream restoration should be a focal point around which this post-industrial transition takes place.

Remote Management

Given the inherent complexities in watershed management, strategies for restoration success may come from a variety of sources. The following section attempts to categorize those efforts that are conducted remotely as opposed to those activities that have an on the ground component.

Governance Challenges and Opportunities

In a society where political decisions change frequently and where regulation tends to be used to achieve environmental improvement, river management has historically been addressed through control and static engineered solutions (e.g., dredging, armoring, berming). Complicating management of the Oiartzun watershed is the fragmentation of both ideologies and responsibilities among numerous parties responsible for watershed improvement (Figure 6; also see Adams and Lawhon, this issue). Despite these challenges a recent Basque government report indicates that the commitment towards meeting EU water policy, combating climate change in regards to water security, and addressing UN Millennium goals of providing safe drinking water and sanitation are being taken seriously (Basque Water Agency, 2006). Still long term ecological restoration will require decisions that make ecologic sense as well as provide benefits for the human community.

The European Union Water Framework Directive (WFD), entered into law in 2000, and established a mandate for community action in the field of water management ("Flood Prevention in the Autonomous Community of the Basque Country, Spain" n.d.). The Spanish Government under direction from the EU has set up basin management zones around Spain. These "hydrographic demarcations" divide the Basque territory in a way which, "causes considerable difficulties insofar as the handling of available information" between provincial governments (Basque Water Agency, 2006).

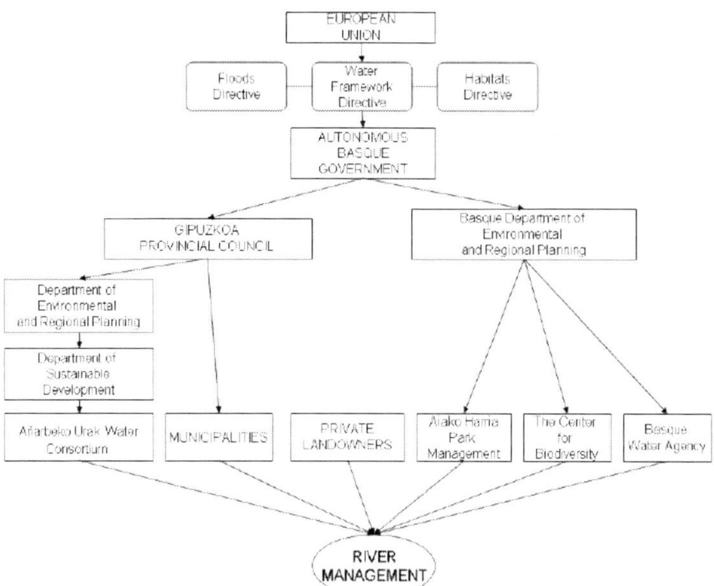

Figure 6. Simplified flow diagram depicting parties responsible for managing rivers in Gipuzkoa Province and their perceived relationship to one in other in terms of management responsibility and institution of oversight.

However, since the Oiartzun River lies entirely within the Basque Autonomous Community, and entirely within one province, Gipuzkoa, the management responsibility defers first to the regional authority, the Basque Water Agency (BWA), an arm of the Basque Department of Environmental and Regional Planning, and also to Añarbeko Urak, a consortium which supplies water to San Sebastian and other municipalities including Errenteria, Hernani, and Oiartzun. The BWA is in charge of planning, constructing, and operating major water infrastructure such as dams, elaborating basin plans, monitoring and enforcing water quality targets, granting permits to use water, undertaking hydrological studies, and providing advisory services to other entities. Provincially, the Gipuzkoa government maintains a hydro-meteorological monitoring network, with twenty-two control stations (one on the Oiartzun) as well as a network of water quality sampling sites in the province (seven on the Oiartzun).

Another driving force influencing watershed initiatives in the region is the EU Habitats Directive. The Habitats Directive of the European Union declares alluvial forests consisting of alder and ash *(Fraxinus excelsior* L.*)* in addition to the mixed riparian woodland on the Atlantic coast as habitats of community interest and to be included in the network of protected areas. Also included on this list are "water courses of plain to montane levels with Ranunculion fluitantis and Callitricho-Batrachion vegetation." According to the LIFE-project 2005 report, "Measures will also be taken to restore fluvial habitats and species, particularly the plant *Soldanella villosa*, given the importance of Aiako Harria [riparian areas] for the global conservation of this species." Thus far Park authorities have established a policy of no-harvesting within a twenty meter buffer on either side of perennial stream courses as well as have banned planting of non-native species on public lands. These measures are a first step but as mentioned previously do not necessarily consist of "best" management especially when, for instance, grazing is still allowed in these areas.

Undetermined as of yet is the impact to watershed management that the 2007 EU Floods Directive will have in the region. The Directive is aimed at assessing, reducing, and managing flood risks in Member States. It requires that preliminary flood risk assessments be carried out by 2011 with plans to prevent, protect, and prepare communities for flood events by 2015. Despite abundant government sponsored literature regarding the sustainable use of water and the protection of river resources, conventional flood control solutions currently remain dominant in the Basque Country. For example, the Department of Planning and Environment of the Basque government has recently appropriated 60 million Euros to construct 18 "preventative" measures to stop flooding in the Urumea River, the watershed directly west of the Oiartzun whose headwaters also are within the Park boundary ("Contenidos - URA - Agencia Vasca del Agua," n.d.). While some of the funding seeks to rebuild undersized bridges (with the secondary effect of benefiting channel stability and stream ecology), most of these proposed projects are aimed at channel dredging to increase channel width and depth, and engineering bypass channels at the downstream end of the river (DHI Group, 2009). These approaches do little to address water quality or habitat. As stated earlier it seems that a coordinated effort to integrate flood reduction efforts with habitat and water quality improvements through riparian reforestation and wetland development would provide multiple benefits for the watershed community while addressing the Water, Habitat, and Flood Directives.

Technical and Economic Opportunities

While habitat improvements are certainly an important benefit of restoring floodplain forests, the other services that these zones provide to the community are multifaceted and often overlooked (Table 2). Addressing riparian restoration in the context of ecosystem services provides a compelling argument beyond their critical habitat value especially when compared to the single-purpose, capital intensive work of river channel control and manipulation.

Determining the extent of ecosystem service benefit is typically not altogether straightforward. One effort has been the development of modeling programs that predict riparian reforestation effects on flooding. Such programming occurred in a multi-country LIFE-Environment Project that focused on understanding how computer modeling techniques could be used to assess floodplain reforestation.

The results indicated that embankments such as flood levees that separated a river from its floodplain could increase flood flows downstream by 50-150%. Restoring the river channel through the floodplain to its pre-engineered dimensions, however, could reduce flood flows 10-15% which may also assist in the restoration of target riparian communities and other secondary benefits.

A separate but similar study on the River Parrett in the United Kingdom revealed comparable results indicating that a forested floodplain increased water storage by 71% over a grassed floodplain and delayed the downstream progression of the flood peak considerably.

Table 2. Summary of riparian ecosystem services and their classification as economic, environmental, and/or social benefits (Thomas and Nisbet, 2007; Smith et al., 2008; National Research Council, 2002)

Example Riparian Forest Ecosystem Services			
Benefit	Economic	Environmental	Social
Biodiversity improvement (of both the terrestrial and aquatic ecosystems)	√	√	√
Improved water quality (including: sediment reduction; reduction of thermal pollution through shading; removal of nitrogen and non-point source pollutants such as heavy metals, hydrocarbons, and pesticides; organic matter processing)	√	√	√
Flood reduction and maintenance of baseflow during droughts	√	√	√
Aesthetics and recreation	√		√
Other agricultural benefits such as pollination services, windbreaks, possible forage	√	√	
Carbon sequestration	√	√	√
Property Damage and Maintenance Reduction	√		√

Such information would be highly valuable when used in a detailed analysis of the entire Oiartzun catchment. Specifically, as suggested by the River Parrett study, from a hydraulic and hydrologic perspective it is important to know where reforestation and development of floodplain wetlands would have the greatest ability to reduce and/or desynchronize the peak flows from individual tributaries in order to minimize downstream peak flood flows (Thomas and Nisbet, 2007).

A green engineering approach to reducing downstream flooding was made famous by the Charles River Natural Valley Storage Project which was carried out as an alternative to large dam building in Massachusetts' Charles River watershed by the Army Corp of Engineers between 1968 and 1972. Floodplain easements (4,860 acres) and wetlands (3,250 acres) acquisitions were accomplished by 1984. Total storage provided by the system was estimated at 50,000 acre feet of water, equivalent to a typical New England reservoir (Riley, 1998). Quoting the Corps' 1972 study report: "Nature has already provided the least-cost solution to future flooding in the form of extensive [riverine] wetlands which moderate extreme highs and lows in streamflow. Rather than attempt to improve on this natural protection mechanism, it is both prudent and economical to leave the hydrologic regime established over millennia undisturbed. In the opinion of the [Army Corps] study team, construction can add nothing" (USACOE, 1972). It should be noted that among the major areas the Corps found critical to control flooding, many were found in tributaries some distance from the mainstem.

Conclusion

It seems that where river restoration efforts are successful, they are usually encapsulated by a river corridor planning process which is a public driven process enacted in order to protect and improve the long term quality and quantity of water as well as provide additional benefits of biodiversity, social, and economic capital. For example, on the Guadajoz River (Spain) a LIFE project has supported development of a sustainable management plan for the basin. “The project has helped the region meet the requirements of the WFD. It has analyzed the environmental problems and devised solutions. Several techniques have been tested including erosion control, improvement of water quality, monitoring, definition of ‘ecological’ flow levels, and restoration of natural vegetation. Effective techniques were also established for regenerating different areas of ‘run down’ or unmanaged riverside habitats” (Jones and Europaische Kommission, 2007)

In the Oiartzun watershed, as with nearly all watersheds, successful on the ground implementation will benefit from stakeholder participation and adaptive management (Eyquem, 2007). The vision for the process embraces new standards for forestry and grazing activities in the Park and asks that managers rethink old ideas about such things as the need for a dense road network and the idea that clearing brush, shrubs, and fallen wood provides a better aesthetic. In the villages, historic features such as the Arditurri Mines should be embraced for their ability to attract tourism yet their legacy of toxicity must be dealt with. So too must be rethought the idea of river control and alteration. Provision of a green corridor that supports wildlife, recreational opportunities, and water purification among other benefits will surely serve the community better in the long term than the highly altered and sometimes fully enclosed drainage courses that today constitute streams. This may even mean that in the heaviest urban areas roads may be rerouted and buildings removed in order to make room for river meanders and flooding. Wetlands also will need to be remembered for their ability to provide microhabitats, landscape diversity, and flood reduction.

The goal of such projects is the reduction of flood peaks, the return of a healthy river ecosystem including perhaps salmon, and the reinvigoration of the human communities that want to gather near clean flowing water. Perhaps Gipuzkoa may even emulate the *Sea to Sky Riparian Corridor (*A multi-town economic and quality of life venture based near the resort town of Whistler, British Columbia, Canada where users can hike or bike from the sea to the mountains) so that ambitious mountaineers could one day leave from Pasaia and hike or bike their way up a trail through floodplain parks and riparian forests eventually reaching the high summits of Aiako Harria Nature Park. So too may Gipuzkoa want to emulate the many cities and towns in the American west whose development of whitewater boating parks for river recreation have had added affects of improving public open space and access to rivers as well as providing a significant economic input to these communities.

In summary, green infrastructure development of a riparian corridor and rehabilitation of floodplain wetlands would provide habitat, flood reduction, and water quality benefits to the Oiartzun River thereby addressing three UN Directives related to river and forest management. The secondary benefits of this green infrastructure would be numerous including social, economic, and ecological services for the entire community. Key to the process is developing longitudinal and lateral connectivity through headwaters, transitional, and depositional riparian areas as well as developing the management capacity and communication tools to effectively implement this endeavor.

References

Ansorena, J., Marino, N. and Legorburu, I. (1995). Agricultural Use of Metal Polluted Soil Near an Old Lead-Zinc Mine in Oiartzun (Basque Country, Spain). *Environmental Technology*, 16(3), 213-222.

Barten, P. K., Kyker-snowman, T. [., Lyons, P. J., Mahlstedt, T., O'Connor, R., Spencer, B. A. (1998). Massachusetts: Managing a Watershed Protection Forest. *Journal of Forestry*, 96, 10-15.

Basque Water Agency. (2006). Water Stewardship in the Autonomous Community of the Basque Country. Basque Water Agency, UNESCO. Retrieved from http://www.unesco.org/water/wwap/wwdr/wwdr3/case_studies/pdf/Water_Stewardship_in_the_Autonomous_Community_of_the_Basque_Country.pdf.

Berg, A. (2002). Breeding birds in short-rotation coppices on farmland in central Sweden--the importance of Salix height and adjacent habitats. *Agriculture, Ecosystems and Environment*, 90(3), 265-276.

Borja, A., Galparsoro, I., Solaun, O., Muxika, I., Tello, E., Uriarte, A. et al. (2006). The European Water Framework Directive and the DPSIR, a methodological approach to assess the risk of failing to achieve good ecological status. *Estuarine, Coastal and Shelf Science*, 66(1-2), 84-96.

Borja, A., Garcia, B., Blanco, J., Agirre, A., Aierbe, E., Bald, J. et al. (2003). Network Monitoring of Surface Water Masses of CAV: Volume 13: Hydrologic Unit of Oiartzun. Department of Land Use Planning and the Environment, Basque Government.

Borja, A., Bald, J., Franco, J., Larreta, J., Muxika, I., Revilla, M. et al. (2009). Using multiple ecosystem components, in assessing ecological status in Spanish (Basque Country) Atlantic marine waters. *Marine Pollution Bulletin*, 59(1-3), 54-64.

Brierley, G. J. and Fryirs, K. A. (2008). *River Futures*. Island Press.

Contenidos - URA - Agencia Vasca del Agua. (n.d.).. Retrieved November 10, 2009, from http://www.uragentzia.euskadi.net/u81-0003/es/contenidos/nota_prensa/inundabilidad_urumea/es_prensa/indice.html.

Destruction at all Co[a]st: 2008: Report on the Spanish Coast Situation. (n.d.) . Greenpeace International.

DHI Group. (2009). Let the river breathe. Retrieved December 8, 2009, from http://www.dhigroup.com/?sc_itemid=%7B8D72AF4D-5E8E-4CAC-AC35-1B8D82416771%7D.

Diez, J. R., Elosegi, A. and Pozo, J. (2001). Woody debris in North Iberian streams: influence of geomorphology, vegetation, and management. *Environmental Management*, 28(5), 687–698.

Docampo, L., Rico, E., Sevillano, M., De Bikuna, B., and Rallo, A. (1991). Morphology and quantitative analysis of fluvial erosion systems in the hydrological network of the Basque Country autonomous region. *Journal of Environmental Management*, 32(3), 251-266.

Elosegi, A., Antón, A., Aristegi, L., and Diez, J. R. (2008). Restoring stream channel complexity with large wood. Effects on ecosystem structure and functioning. *Proceedings 4th ECRR Conference on River Restoration.*

EUROPA - Life Projects - Project Summary. (n.d.). *Living River Liesing - Demonstrative Ecological Reconstruction of a Heavily Modified Waterbody in an Urban Environment.* Retrieved January 15, 2010, from http://ec.europa.eu/environment/life/project/Projects/index.cfm?fuseaction =home.createPageands_ref=LIFE02%20ENV/A/000282.

Eyquem, J. (2007). Using fluvial geomorphology to inform integrated river basin management. *Water and Environment Journal*, 21(1), 54-60.

FISRWG. (1997). *Stream corridor restoration handbook an interagency partnership.* Natural Resources Conservation Service.

Flood Prevention in the Autonomous Community of the Basque Country, Spain - WaterWiki.net. (n.d.). Retrieved October 8, 2009, from http://waterwiki.net/index.php/Flood_Prevention_in_the_Autonomous_Co mmunity_of_the_Basque_Country,_Spain.

Giuliano, W. M. and Homyack, J. D. (2004). Short-Term Grazing Exclusion Effects on Riparian Small Mammal Communities. *Journal of Range Management*, 57(4), 346-350.

Gordon, A. M. and Newman, S. M. (1997). *Temperate Agroforestry Systems.* CABI.

Jones, J. A., Swanson, F. J., Wemple, B. C., and Snyder, K. U. (2000). Effects of Roads on Hydrology, Geomorphology, and Disturbance Patches in Stream Networks. *Conservation Biology*, 14(1), 76-85.

Jones, W. and Europaische Kommission. (2007). *LIFE and Europe's rivers : protecting and improving our water resources.* Luxembourg: Office for Official Publ. of the Europ. Communities.

Klapproth, J. C. and Johnson, J. E. (2001). Understanding the Science Behind Riparian Forest Buffers: Planning, Establishment and Maintenance. Virginia Cooperative Extension. Retrieved from www.ext.vt.edu/pubs/forestry/420-155/420-155.html.

Langer, E. L., Steward, G. and Kimberley, M. (2008). Vegetation structure, composition and effect of pine plantation harvesting on riparian buffers in New Zealand. *Forest Ecology and Management*, 256(5), 949-957.

LIFE Guadiato Report. (n.d.). Retrieved from http://ec.europa.eu/ environment/life/project/Projects/files/laymanReport/LIFE00_ENV_E_00 0425_LAYMAN.pdf.

Meyer, J. L., Strayer, D. L., Wallace, J. B., Eggert, S. L., Helfman, G. S., and Leonard, N. E. (2007). The contribution of headwater streams to biodiversity in river networks. *JAWRA Journal of the American Water Resources Association*, 43(1), 86-103.

Montgomery, D. R. and Buffington, J. M. (1997). Channel-reach morphology in mountain drainage basins. *Geological Society of America Bulletin*, 109(5), 596-611.

Naiman, R. and Décamps, H. (1997). The ecology of interfaces: Riparian Zones. *Annual Review of Ecology and Systematics*, 28(1), 658, 621.

National Research Council. (2002). *Riparian Areas: Functions and Strategies for Management*. Washington, D.C: National Academy Press.

Olarieta, J. R., Besga, G.,Virgel, S. (1999). Sediment enrichment ratios after mechanical site preparation for Pinus radiata plantation in the Basque Country. *Geoderma*, 93(3-4), 255–267.

P. Departmento de Medio Ambiente, Planificacion Territorial, Agricultura y Pesca. (2006). Estudio de caso: País Vasco; Un estudio de caso del WWAP preparado para el 2 Informe de las Naciones Unidas sobre el Desarrollo de los Recursos Hídricos en el Mundo: "El agua, una responsabilidad compartida".

Palmeirim, J. M. and Hoffmann, R. S. (1983). Galemys pyrenaicus. *Mammalian Species*, (207), 1-5.

Parendes, L. A. and Jones, J. A. (2000). Role of Light Availability and Dispersal in Exotic Plant Invasion along Roads and Streams in the H. J. Andrews Experimental Forest, Oregon. *Conservation Biology*, 14(1), 64-75.

Ramirez, L. (2005). Oiartzun River: From the mountain to the harbor. Retrieved from http://www.laanunciataikerketa.com/montana_publi.htm.

Remondo, J., Soto, J., González-D'iez, A., D'iaz de Terán, J. R., and Cendrero, A. (2005). Human impact on geomorphic processes and hazards in mountain areas in northern Spain. *Geomorphology*, 66(1-4), 69–84.

Riley, A. L. (1998). *Restoring Streams in Cities: A Guide for Planners, Policymakers, and Citizens* (1st ed.). Island Press.

Rosgen, D. (1996). *Applied River Morphology* (2nd ed.). Wildland Hydrology.

Rupprecht, R., Kilgore, C. and Gunther, R. (2009). Riparian and Wetland Buffers for Water-Quality Protection: A review of current literature. *Stormwater*, 10(8), 46-51.

Schumm, S. A. (1977). *The Fluvial System*. The Blackburn Press.

Smith, M., Schiff, R., Olivero, A., and MacBroom, J. (2008). The Active River Area: A conservation framework for protecting rivers and streams. The Nature Conservancy. Retrieved from floods.org/PDF/ASFPM_TNC_Active_River_%20Area.pdf.

Strahler, A. N. (1952). Hypsometric (area-altitude) analysis of erosional topography. *Geological Society of America Bulletin*, 63(11), 1117.

Sweeney, B. W. (2004). Riparian deforestation, stream narrowing, and loss of stream ecosystem services. *Proceedings of the National Academy of Sciences*, 101(39), 14132-14137.

Thomas, H. and Nisbet, T. R. (2007). An assessment of the impact of floodplain woodland on flood flows. *Water and Environment Journal*, 21(2), 114-126.

Uriarte, A. and Borja, A. (2009). Assessing fish quality status in transitional waters, within the European Water Framework Directive: Setting boundary classes and responding to anthropogenic pressures. *Estuarine, Coastal and Shelf Science*, 82(2), 214-224.

U.S. Army Corps of Engineers. 1972. *Charles River Massachusetts Study Main Report and Attachments.* Department of the Army. New England Division. Corps of Engineers. Waltham, MA.

Vannote, R., Minshall, G., Cummins, K., Sedell, J., and Cushing, C. (1980). The river continuum concept. *Canadian Journal of Fisheries and Aquatic Sciences*, 37(1), 137, 130.

Verry, E. S., Hornbeck, J. W. and Dolloff, C. A. (1999). *Riparian Management in Forests of the Continental Eastern United States* (1st ed.). CRC Press.

Weller, D. E., Jordan, T. E. and Correll, D. L. (1998). Heuristic models for material discharge from landscapes with riparian buffers. *Ecological Applications*, 8(4), 1156-1169.

Zabala, J., Zuberogoitia, I. and Martínez-Climent, J. A. (2006). Factors affecting occupancy by the European mink in south-western Europe. *Mammalia*, 70(3-4), 193-201.

In: Multiple-Use Management ... ISBN 978-1-61942-328-2
Eds: A. DuVal, T. Muruetagoiena et al.

Chapter 3

CHANGING VALUES, CHANGING FORESTS: CONVERTING RADIATA PINE PLANTATIONS AND RESTORING NATIVE FOREST IN THE BASQUE COUNTRY

Peter J. Caligiuri
Yale School of Forestry and Environmental Studies, New Haven
The Nature Conservancy in collaboration with the
US Forest Service, Oregon, US

ABSTRACT

Single-species, even-age plantations of Monterey pine (Pinus radiata D. Don.) are ubiquitous across the landscape of the Basque Country in Northern Spain. These plantations were planted primarily during the twentieth century for timber production, site stabilization, and land rehabilitation. However, recent shifts in the political, economic, social, and ecological values governing native forests underscore the need for a commensurate shift in forest management. This new paradigm would aim to convert existing marginally valuable, single-species plantations to native, mixed-species, deciduous forests in order to capture the myriad goods and services native forest ecosystems provide. The present analysis outlines the major principles of forest dynamics and forest degradation as they apply to public forest management in the Basque Country, with Aiako Harria Nature Park serving as a reference case. The analysis draws on native forest restoration literature from temperate and tropical regions alike to illustrate a restoration pathway that utilizes exotic species to facilitate natural succession of native forest on degraded lands. This evidence supports the conclusion that plantation conversion is a cost-effective and ecologically sound method of accelerating understory re-initiation and succession to species-rich native forest while sustaining an economically viable forest products industry.

Keywords: Basque Country, northern Spain, Aiako Harria Nature Park, *Pinus radiata*, silviculture, forest restoration, forest management, forest stand dynamics, forest degradation

INTRODUCTION

The history of land-use change in the Basque Country is as old as the civilization that has inhabited the mountainous region of northern Spain for millennia. Palynological records from the Cantabrian range in northwest Spain indicate significant decreases in forest cover due to anthropogenic activity as early as 3000 BP (García *et al.*, 2005). Over the past several centuries native forest loss globally and locally has accelerated with the increasing demands of growing populations for food, fuel, timber, and fiber from forests (see e.g., Fimbel and Fimbel, 1996; Geldenhuys, 1997; Ashton *et al.*, 1998; Kenk and Guehne, 2001; Zerbe, 2002).

Across Europe extensive replacement of native forests with agricultural and grazing lands occurred throughout the seventeenth, eighteenth, and nineteenth centuries. At the same time, extensive livestock grazing within forests coupled with intensive timber extraction for shipbuilding, construction, and charcoaling exacerbated the decline of the remaining native forests (Kenk and Guehne, 2001; Zerbe, 2002; García *et al.*, 2005; Onaindia and Mitxelena, 2009). Anthropogenic disturbance in central Europe eventually led to the loss of fifty percent of total native forest area (Kenk and Guehne, 2001).

Native forest loss in Spain during the same period followed a similar pattern (Pardo *et al.*, 2004; García *et al.*, 2005; Onaindia and Mitxelena, 2009). Then, beginning at the turn of the twentieth-century, traditional land uses in the Basque Country were gradually supplanted by plantation forestry with exotic conifers, primarily Monterey pine (*Pinus radiata* D. Don.) (Schmitz *et al.*, 1998). As a result, the contemporary forest landscape is a highly fragmented matrix of pine plantations with small and often isolated stands of native forest dispersed throughout (Rescia *et al.*, 1997).

However, just as the political, social, and economic drivers of deforestation and reforestation with exotic species have changed over time, so too do the political, social, and economic values that govern forest management and restoration (García Pérez and Groome, 2000). As these values change, paradigms for forest management must evolve to meet new objectives. Empirical evidence gathered during a ten-day research trip to Aiako Harria Nature Park in October 2009 suggests that growing awareness of the importance of functioning native forest ecosystems within Basque society is currently driving such a change. Recent attempts to integrate ecological principles into forest management practices (especially on public lands) through the Basque Country Forest Plan support this conclusion (Montiel and Galiana, 2005). Success in this endeavor hinges upon striking a balance between native forest restoration and production forestry, which in turn requires a sound understanding of native forest ecology and restoration.

The analysis presented herein begins with a discussion of historical land-use patterns and the current context of forest management in the Basque Country. It then turns to the principles of forest dynamics and forest degradation as they apply to natural succession and restoration of moist, temperate native forests like those of Aiako Harria Nature Park. The discussion concludes by outlining the general silvicultural components of an appropriate restoration pathway that would promote a more resilient and productive forested landscape.

LAND-USE CHANGE AND FORESTRY IN THE 20TH CENTURY

Agricultural production in the Basque Country reached its maximum extent near the end of the nineteenth century as population in the region peaked. At that time, cultivated land under production (primarily cereal crops for human consumption) in the Pyrenean region of Spain occupied more than a quarter of the area below 1600 meters (García-Ruiz *et al.*, 1996). Schmitz (1998, p. 137) characterized the nineteenth century Basque landscape as a "...heterogeneous mosaic of pastures and crops with hedgerows and native deciduous forests."

However, by the turn of the twentieth century a series of socio-economic forces combined to create a shift away from agricultural land use. Increasing urbanization and industrialization led to a breakdown of the traditional social order and a marginalization of Basque agrarian society, which in turn sparked a decline of the rural, agricultural population (García-Ruiz *et al.*, 1996; Rescia *et al.*, 1997; Schmitz, 1998). These factors contributed to the evolution of land-use and forest composition patterns in the region. García-Ruiz *et al.* (1996) outlined the following changes: i) a decline of sheep and an increase in cattle production, ii) a decline in the total cultivated area, iii) a decline in cereal production and increase in abandoned cropland conversion to meadow and scrubland, iv) a decline in native deciduous forests, and iv) an increase in reforestation of abandoned agricultural lands with exotic conifers, predominantly Monterey pine.

Widespread reforestation with exotic conifers began in the late 1800's and early 1900's as Spanish foresters were called upon to provide a technical solution to curb rampant deforestation and stabilize abandoned agricultural lands (García Pérez and Groome, 2000). Simultaneously, Monterey pine was gaining worldwide recognition as a fast growing, hardy plantation species capable of producing valuable wood and fiber in a variety of climates and growing conditions. Driven by technocratic forest policy and economic interests, the total area of Monterey pine plantations in northern Spain increased dramatically during the first half of the twentieth century (Scott, 1960; García Pérez and Groome, 2000; Onaindia and Mitxelena, 2009). By 1960, 51,000 hectares of Monterey pine were planted across northern Spain, the highest concentration of which was in the Basque Country (Scott, 1960). The shift towards the modern Basque forest landscape was underway.

In 1996, total forested area in the Basque Country (defined as canopy coverage greater than 10%) was 390,005 hectares, equal to fifty-four percent of the total land area (National Forest Inventory, 1996). According to Spain's 1996 National Forest Inventory (IFN) data 150,199 hectares, or thirty-nine percent of the total forested area, was planted in Monterey pine (Table 1). In the provinces of Vizcaya and Guipuzcoa the percentage of Monterey pine plantation forest cover is higher still, surpassing sixty-two and forty-six percent of total forested area, respectively (National Forest Inventory, 1996). In other words, while total forested area increased substantially during the later half of the twentieth century, it was primarily the result of afforestation in plantations of Monterey pine. Native forest loss and fragmentation continues today within a landscape dominated by homogenous, single-species, even-age conifer plantations (Onaindia and Mitxelena, 2009).

Table 1. Forested area occupied by Pinus radiate D. Don. in the Basque Country, Spain

Province	Alava	Vizcaya	Guipuzcoa	Total
Total Land Area (ha)	304.700	221.700	199.700	726.100
Total Forest Area 1996 (ha)	143.506	128.244	118.255	390.005
Percent Forested Area	47%	58%	59%	54%
Pinus radiata 1996 (ha)	15.678	79.726	54.795	150.199
Percent Forested Area in *Pinus radiata*	11%	62%	46%	39%

Source: Inventario Forestal Nacional, 1996; Chauchard, 2000.

Forest management in the Basque Country is further complicated by the distribution and size of plantation ownerships. For example, the roughly 150,000 hectares of Monterey pine plantation are divided among 25,000, mostly private, plantation owners. Average plantation size is approximately six hectares, yet seventy-three percent are less then five hectares and thirty percent are between two and five hectares (Schmitz *et al.*, 1998; Chauchard, 2000). Chauchard (2000) explains that the small size and high number of private and public plantation forests creates a communication barrier between forest owners and the forest products industry. Lack of forest management consultation translates into neglect of important silvicultural treatments that would otherwise optimize timber growth, yield and quality. As a result, plantation management in the Basque Country is often low intensive with expected implications on the health and productivity of these forests.

Presently, global and regional softwood supply is more than sufficient to meet market demand. Therefore, plantation forestry investments are financially viable if and when suppliers can realize some competitive advantage in the market (e.g., low planting and harvesting costs, low transportation costs, and/or high quality common woods) (Oliver and Meznick, 2005). In these terms, the competitive advantage of the Basque forest sector is slim. Considering both the weak market position and the low timber value of Basque Monterey pine plantation forestry, decision-makers within government and industry would be wise to reevaluate the guiding forest management paradigm, including strategies to diversify production, increase cost-effectiveness, and create added-value (Euskadi Forestal, 2005).

Modern Forest Management: Challenges and Opportunities

The Basque Country faces a challenging natural resource management dilemma with significant social, ecological, and economic implications. In light of i) the changing values and objectives governing forest management, ii) the marginal value and productivity of Monterey pine plantations, and iii) the current softwood market dynamics, a series of questions emerge. What alternatives strategies exist to manage the vast area of even-aged, low-value Monterey pine plantation forests? More specifically, how can the forest management paradigm evolve in response to changing objectives calling for increased resiliency, stability, and provision of a wider array of forest ecosystem goods and services? As Parrotta *et al.* (1997) explains, the answer to these questions ultimately depends on the objectives of the stakeholders, the costs and benefits of possible management strategies, and the social, ecological, and economic values of the forestland in its present versus desired condition.

This is not to imply that the forest sector in the Basque Country is superfluous. On the contrary, the forest products industry appears to be a small but central component of the regional economy. However, because intensive forest management necessarily involves higher silvicultural costs to optimize timber growth, yield, and quality, these investments should be concentrated on sites with the greatest potential returns (Yin and Sedjo, 2001; Allen *et al.*, 2005).

Therefore, this analysis calls for a holistic forest management strategy that focuses timber production on the most productive (often private) forestland capable of generating higher volumes of high quality timber. This strategy in turn reduces the need for plantation forestry on ecologically and culturally important (often public) forestlands, allowing forest managers to prioritize other objectives, such as biodiversity conservation, watershed protection, and cultural legacy, when restoring and managing public forestlands like Aiako Harria Nature Park. This may represent a politically feasible, cost-effective, and ecologically sound solution to the present dilemma. The following section outlines historical evidence in support of this alternative strategy for public forest management in the Basque Country.

OLD-FIELD PINE ANALOGS: THE PRECEDENT FOR RESTORATION

There is a good deal of research highlighting the negative aspects of homogenous, single-species, even-age plantations, including higher rates of water use compared to natural vegetation, increased erosion from intensive management, decreased biodiversity, and instability/susceptibility of the stand to disturbance agents. These criticisms are used to argue for the replacement of plantation forests with native forestland. However, many of the same studies acknowledge the integral role plantation forests can play in facilitating the rehabilitation and natural succession of mixed-species, native forests on degraded agricultural lands (see e.g., Geldenhuys, 1997; Schmitz *et al.*, 1998; Kenk and Guehne, 2001; Onaindia and Mitxelena, 2009). In this case, the history of old-field succession in the eastern United States is an insightful analog of relay floristics and secondary succession following the establishment and growth a single-species, even-age forest.

The history of native forest loss, agricultural land-use, and subsequent abandonment in Northeastern United States is strikingly similar to the Basque case. By the late 1800's between sixty and eighty percent of the northeastern forests were cleared for agricultural production and pastureland. With the settlement of the Ohio Valley and the social and economic turmoil created by the Civil War much of this land was abandoned by the end of the nineteenth century. Agricultural abandonment set in motion a process of succession in which these old fields were colonized by natural regeneration of wind-dispersed seeds from tolerant conifers like eastern red cedar (*Juniperus virginiana* L.) and eastern white pine (*Pinus strobus* L.) (Raup, 1966; Foster, 1992; Howard and Lee, 2002).

Over the next half-century, eastern white pine grew into the dominant forest species of the Northeastern landscape. Through a process known as relay floristics, this single species altered environmental conditions (i.e., by providing shade, increasing soil humus, soil moisture, and nutrients) and facilitated the regeneration of subsequent species (see e.g., Clements, 1916). The structural and functional attributes that accrued over the intervening decades helped to ameliorate the harsh open field microclimate, providing suitable regeneration microsites for less tolerant, animal-dispersed hardwoods (Howard and Lee, 2002; Kirkman *et al.*, 2007). As the white pine matured, harvesting and natural disturbance gradually removed the overstory, releasing advanced regeneration of native hardwoods in the understory. The eventual result was a second-growth, even-aged, mixed-species hardwood forest standing in its place (Raup, 1966; Foster and Boose, 1992; Barten *et al.*, 1998; Howard and Lee, 2002).

The Piedmont region of the Southeastern United States provides a contrasting but nonetheless informative old-field analog relevant to the Basque case. Whereas in the Northeast the pine overstory was removed allowing secondary succession of native hardwood forest, in the Southeast many agricultural lands were intentionally planted and managed for timber production of native softwoods, most often loblolly pine (*Pinus taeda* L.) (Felix *et al.*, 1983). In the most productive stands, additional management investments, including control of competing hardwoods, fertilizer inputs, and artificial regeneration, has precluded secondary succession and led to high timber growth and yield.

These two examples of old-field pine succession are useful to this discussion of forest management and restoration. On the one hand, the successional processes at work in the transition from a single-species, eastern white pine to mixed hardwood forest in the Northeast illustrates the potential utility of radiata pine plantations in the restoration of mixed-species forests in the Basque Country. Conversely, the establishment of highly productive but intensively managed plantation forests in the Southeast underscores the importance of concentrating plantation forestry where increased management costs are justified by higher growth and yield (Yin and Sedjo, 2001; Allen *et al.*, 2005). Together these examples highlight alternative approaches to managing what are, in practical terms, "old-field" Monterey pine forests in the Basque Country. Assuming that economic forces will concentrate plantation forestry on the most productive and economically viable sites in Basque Country, this analysis now turns to the question of where and how to implement native forest restoration across the forested landscape.

NATIVE FOREST RESTORATION

Potential restoration pathways range from one-time, total liquidation of the existing plantation forests followed by replanting with the desired mix of native forest species – what Parrotta *et al.* (1997) refers to as full ecosystem restoration – to the gradual conversion of plantation forests through the use of non-native species as a functional bridge for advanced regeneration (Ashton *et al.*, 1997; Kirkman *et al.*, 2007). The second option, like the old-field analog presented above, utilizes less intensive silvicultural treatments like single-tree, group, and patch selection to gradually remove the overstory, promote natural regeneration, and facilitate secondary succession (Mosandl and Kleinert, 1998; Mosandl and Küssner, 1999; Kenk and Guehne, 2001).

In many situations the choice between restoration pathways (and between forest restoration and plantation forestry) depends on the tradeoffs between economic feasibility, site suitability/productivity, and ecological integrity. For example, full ecosystem restoration of a diverse native forest vis-à-vis mixed plantings and intensive site management may result in favorable biodiversity recovery. However, the high cost of large-scale application of this option may make it economically unfeasible (Parrotta *et al.*, 1997). Consequently, a growing area of forest restoration research is examining alternative pathways that integrate natural processes (i.e., natural regeneration, understory re-initiation, and secondary succession) to achieve a less intensive, more cost-effective means of converting plantations to native forestland (Ashton *et al.*, 2001; Kenk and Guehne, 2001; Zerbe, 2002).

It is important to recognize that *how* native forests are restored and the success of restoration efforts depends upon forest managers' understanding of *what* they are restoring for and *why* (Ashton *et al.*, 2001). This requires a comprehensive understanding of the social values that drive land and forest use, which is only possible through stakeholder involvement in a participatory approach to land-use decision-making (see Adams and Lawhon, this volume). In addition, successful restoration strategies must be based on knowledge of forest dynamics and forest degradation, which together drive regeneration, succession, and forest development spatially and temporally. The following sections will use Aiako Harria Nature Park as a reference to illustrate these concepts as they apply to native forest restoration in the Basque Country.

FOREST DYNAMICS

Four general principles of forest dynamics are important to the restoration of native temperate, moist forests of the Basque Country. They include i) the relationship between succession and natural disturbance regimes, ii) the link between autecology, site restrictions, and forest composition, iii) the role of species guilds in creating forest structure, complexity, and function, and iv) the influence of forest stratification processes on forest structure, complexity, and function. These principles will guide the subsequent discussion of alternative silvicultural systems for native forest restoration in the Basque Country.

First, succession of native forests is intimately related to natural disturbance regimes. Disturbances open growing space for the initiation and/or release of regeneration from newly colonizing seed, buried seed bank, and advanced regeneration. However, natural disturbance regimes are not monotypic, but vary in type, frequency, magnitude, and size across the landscape. This is a critical factor in the successful regeneration of a diverse but site appropriate suite of tree, shrub, and herbaceous forest species, many of which are dependent on the disturbance regime of a specific site (Oliver and Larson, 1996; Ashton *et al.*, 2001; Kirkman *et al.*, 2007).

In Aiako Harria Nature Park (and the coastal zone of the Basque Country in general), the natural disturbance regime varies across topographic, elevational, and aspect gradients (Figure 1). For example, hotter, xeric conditions and shallower soils are typical at higher elevations along spur ridges and southern aspects. Consequently, these sites are more susceptible to widespread and more frequent disturbances of low and moderate intensity, such as fire and convectional windstorms. Regeneration is primarily early-seral, shade intolerant species that rely on buried seed bank, colonizing seed, or root/stump sprouting. These species are able to germinate or establish on bare mineral soil and grow in the harsh microclimate found on exposed, often south-facing sites. Conversely, cooler, mesic conditions and moderate to deep soils common on mid-slope sites are prone to small, frequent but typically non-lethal disturbances like wind throw, pests, and pathogens. Regeneration on these sites is primarily by mid- to late-successional, shade-tolerant species through advanced regeneration and vegetative propagation, with the exception of limited seed colonization where tip-up mounds and canopy gaps expose bare mineral soil and allow sufficient light penetration for regeneration of early-successional species. More severe but less frequent disturbances like flooding and landslides on mid-slopes and along drainages also occur, creating larger and more variable disturbed areas suitable for regeneration by a mix of early-, mid-, and late-successional species.

Understanding this relationship between disturbance and regeneration is essential for securing the desired and appropriate mix of species for a given site. Through the deliberate and strategic use of natural disturbance imitations vis-à-vis the application of silvicultural treatments, forest managers can restore a complex, mixed-species native forest across the landscape. Forest managers should be aware though that intentional or inadvertent exclusion of natural disturbance processes will likely shift species composition and succession in unknown directions, potentially leading to restoration failure over time.

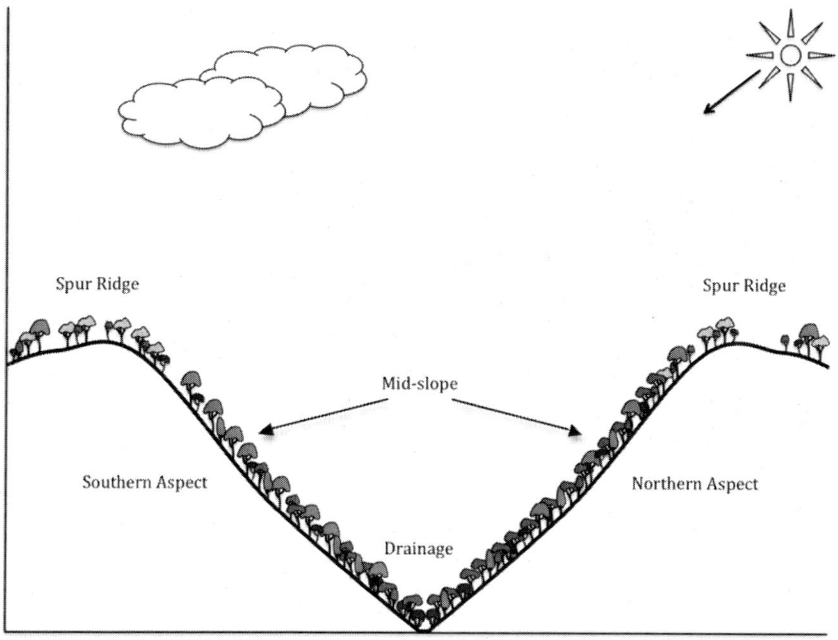

Figure 1. Natural disturbance regimes in Aiako Harria Natural Park vary across the landscape. At higher elevation spur ridges and southern aspects are hotter and more xeric sites, making these areas more prone to larger-scale, frequent, low to moderate severity disturbances, such as fire and convectional windstorms. Conversely, mid-slopes, especially on northern aspects, are typically cooler, more mesic sites prone to small, frequent but non-lethal disturbances like wind throw, pests, and pathogens. Finally, drainages with deep, mesic and hydric soils are often subject to less frequent but more severe disturbances, such as flooding and landslides. Forest type, composition, structure, and function, are a reflection of site characteristics. For example, mixed oak forest predominate on hotter, drier southern aspects, while mixed beech forests with their associated shade-tolerant species occupy cooler, more mesic northern aspects. Both mixed oak and mixed beech forests transition to a dispersed oak grassland forest type on spur ridges and exposed, south-facing upper slopes where shallow, xeric soils preclude the establishment of many species.

Second, the major tree species of the forest type determine the basic structure and function of the native forest. Therefore, species autecology provides a valuable map for the progression and eventual outcome of the restoration pathway.

Understanding the degree to which species composition is driven by biophysical site characteristics like topography, elevation, aspect, light, soil moisture, and nutrient availability, is essential to successful restoration of native forest (Oliver and Larson, 1996; Ashton *et al.*, 2001).

The moist, temperate forests of the Basque Country are primarily mixed oak and mixed beech forest types. The oak forest type is composed of canopy dominants *Quercus robur* L. and *Fraxinus excelsior* L., with co-dominants *Castanea sativa* Mill. and *Acer campestre* L.. The beech forest type is dominated by *Fagus sylvatica* L. with dispersed *Quercus robur* and *Ulmus glabra* Huds.

Both forest types contain a mix of sub-canopy and understory species, including *Betula pendula* Roth, *Salix atrocinerea* Brot., *Ulmus minor* L., *Crataegus monogyna* Jacq., *Corylus avellana* L., *Laurus nobilis* L., and *Rhamnus frangula* L.. In the case of the beech forest type the heavy shade of the beech canopy precludes the presence of the more shade intolerant of these species except where gaps in the overstory occur (Onaindia and Mitxelena, 2009). *Malus sylvestris* (L.) Mill., *Prunus spinosa* L., and *Sorbus aucuparia* L. are also common to these forest types and provide an important food source for many wildlife species, whose movement within the forest provides a critical mechanism for seed dispersal.

Empirical evidence from Aiako Harria Nature Park illustrates the concept of autecological differentiation along topographic, aspect, and elevational gradients (Figure 1). For example, mixed oaked forests predominate on warmer, drier, southern aspects. Conversely, closed, mixed beech forests occupy cooler, mesic, northern aspects along drainages, toe-slopes, and upper mid-slopes.

The structure of both forest types becomes increasingly open moving from lower mid-slope to spur ridge. Mixed oak and mixed beech forests transition to a dispersed *Quercus pyrenaica* L. forest type with a dense grass understory on exposed, south-facing upper slopes and spur ridges with shallow, xeric soils. Understanding the relationship between species autecology and site underscores the importance of restoring the appropriate forest type on its appropriate site within the forested landscape.

Third, complex native forest is composed of multiple and distinct sub-groups of species, often referred to as guilds. Guilds are classified according to regeneration origin (i.e., advance regeneration, vegetative reproduction, buried seed bank, or colonizing seed), stage of stand development in which they dominate (i.e., stand initiation, stem exclusion, understory re-initiation or late-successional), and growth habit within the forest strata (understory, sub-canopy or canopy).

Species guilds are the building blocks of native forest, providing the structural and functional complexity necessary to re-initiate successional processes and create resilient, restored forest conditions. The principles discussed above, such as natural disturbance, species autecology, and dispersal mechanism are important factors to consider in the regeneration of multiple guilds (Oliver and Larson, 1996; Ashton *et al.*, 2001).

According to the classification system described above, the native forest tree species of the Basque Country can be grouped into six guilds (Table 2). They are i) pioneers that dominate stand initiation (e.g., *Salix* and *Sorbus*), ii) pioneers that dominate stand stem exclusion (e.g., *Betula* and *Prunus*), iii) late-successional canopy dominants (e.g., *Fagus*, *Quercus*, and *Castanea*), iv) late-successional canopy non-dominants (e.g., *Acer* and *Fraxinus*), v) late-successional subcanopy (e.g., *Ilex* and *Laurus*), and vi) late-successional understory (e.g., *Corylus* and *Rhamnus*).

The fourth and final principle is forest stratification, which occurs in two dominant processes. The first is referred to as dynamic stratification, whereby species guilds of different developmental stages sequentially gain and loose dominance in the canopy. Conceptually, this can be thought of as forest succession over time. In the Basque forest, pioneers of stand initiation and stem exclusion (e.g., *Betula*, Prunus, Salix, and *Sorbus*) colonize and regenerate quickly following disturbance by means of small wind- or animal-dispersed seed but are also generally short-lived. Consequently, when these species die they relinquish growing space to mid- and late-successional species (e.g., *Castanea*, *Fagus*, *Fraxinus*, *Quercus*, and *Ulmus*), which grow to assume dominant and co-dominant positions in the forest canopy (Oliver and Larson, 1996; Ashton *et al.*, 2001).

The second process is referred to as static stratification, whereby species guilds of different growth habits develop simultaneously but differentiate into forest strata, occupying distinct positions within the vertical structure of the forest.

Late-successional Basque forests will stratify according to growth habit and shade tolerance into canopy dominants (e.g., *Castanea*, *Fagus*, and *Quercus*), co-dominants (e.g., *Acer* and *Fraxinus*), and sub-canopy and understory (e.g., *Corylus*, *Ilex*, and *Laurus*) strata. Dynamic and static stratification processes are essential to the development of structural and functional complexity, and hence resiliency within native forests (Oliver and Larson, 1996; Ashton *et al.*, 2001).

Table 2. The native forest species of Aiako Harria Nature Park can be grouped into the six species guilds described above. Guilds are defined according to i) regeneration origin, ii) stage of stand development in which they dominate, and iii) growth habit within the forest strata. Species guilds form the building blocks of native forests, providing the structural and functional complexity necessary to re-initiate successional processes and create resilient, restored forest conditions

Autecology of principle tree species in moist, temperature mixed oak and mixed beech forest - Aiako Harria Natural Park			
Regenerational Guild	Mode of Dispersal/Regeneration	Life History Traits	Disturbance Type
Pioneers of stand initiation *(Salix, Sorbus)*	Wind, Water, Birds	Abundant, Small-seed; Shade-intolerant; Fast-growing; Short-lived; Early successional	Frequent, severe, exposing mineral soil (Landslides, floods, fires, tip-up mounds in large canopy openings
Pioneers of stem exclusion *(Betula, Malus, Prunus)*	Wind, Water, Birds, Buried seed	Abundant, small seed; Shade-intolerant; Fast-growing; Medium-lived; Mid succesional	Same as above but can establish in smaller camopy openings
Late-succesional dominants *(Castanea, Fagus, Quercus)*	Gravity, hoarding mammals, advance regeneration,(Seedlings, saplings, sprouts)	Large seed; Germinates sporadically with mast years; Intermediate –to-shade-tolerant, Long-lived	Non-lethal disturbances promoting release of advance regeneration (Single/multiple tree windthrow, pathogens, low-intensity fire)
Late-succesional co-dominants *(Acer, Fraxinus, Ulmus)*	As above, except wind also carries winged seed	Medium-seed; Germinates seasonally; Intermediate- to shade-tolerant; Long-lived	Non-lethal release disturbances, tip-up mounds in small/medium canopy openings
Late-succesional subcanopy *(Crateagus, Ilex,Laurus)*	Birds, Rodents, Vegetative propogation	Small-seed; Shade-intolerant; Long-lived	Non-lethal release disturbances, vegetative sprouting where stem/roots survive
Late-succesional understory *(Corylus, Rhamnus)*	As above	Medium/small seed; Shade-intolerant; Long-lived	As above

Modified from: Smith et al.,1997; Ashton et al. 2001.

Forest Degradation

Forest degradation describes a set of processes that disrupt the forest dynamic principles outlined above, thereby inhibiting, and in some cases preventing entirely, understory re-initiation, succession, and stratification within forests. Knowledge of these processes and how to correct them is a fundamental component of appropriate silvicultural prescriptions within any restoration pathway (Ashton *et al.*, 2001). There are two distinct types of forest degradation contributing to the current structure, composition, and development of native and plantation forests in the Basque Country.

The first is acute degradation. This refers to one-time disturbances, usually the temporary or permanent conversion of forest to other uses, which have dramatic and long-lasting effects on forest structure, composition, and succession. For example, forest clearing for agriculture and pastureland during the eighteenth and nineteenth centuries in the Basque Country caused permanent, acute degradation of native forest. The scale, intensity, and duration of land clearance are the primary factors affecting the capacity of the forest to recover post-disturbance.

Not surprisingly, permanent conversion is the most detrimental because of the complete removal of all potential regeneration sources (e.g., buried seed bank, advance regeneration, root stock, and stumps), as well as the increased rates of soil erosion and nutrient loss (Schmitz *et al.*, 1998; Ashton *et al.*, 2001). In such cases, reforestation with fast-growing, tolerant tree species (like Monterey pine) is often necessary to reverse acute degradation processes, stabilize degraded lands, and reinstate natural succession; a pattern similar to the northeastern U.S. old-field pine example outlined previously (Parrotta *et al.*, 1997).

The second process is chronic degradation, which refers to disturbances that are more or less continuous. This can occur through i) top-down impacts, such as frequent and repeated removal of vigorous canopy trees through improper silvicultural treatments, or ii) bottom-up impacts, such as the continuous eradication of natural regeneration in the understory by herbivory, understory cultivation, or vegetation control. In either case, the effects are sometimes less noticeable than acute degradation at the outset but the long-term consequences are equally damaging (Ashton *et al.*, 2001).

In the Basque Country, repeated removal of the Monterey pine overstory, followed by replanting, weed control, and pre-commercial thinning contributes to both top-down and bottom-up chronic degradation. Likewise, herbivory by free-range livestock grazing in both plantation and native forests creates bottom-up impacts by eliminating advanced regeneration of mid- and late-successional tree species, as well as other important shade-tolerant understory species. Together, these chronic disturbances severely alter forest dynamics processes, namely, understory re-initiation, secondary succession, and stratification (Ashton *et al.*, 2001). This leads to gradual structural and functional degradation and a decrease in overall forest complexity, which will persist until livestock herbivory is controlled and intensive plantation management is concentrated on production forestlands (Figure 2).

Figure 2. Example of the effects of top-down and bottom-up impacts of harvesting and constant grazing pressure contributing to chronic disturbance within Monterey pine plantations in the Aiako Harria Nature Park. Notice the grass-dominated understory with little or no advance regeneration of native tree species.

Radiata Pine as a Catalyst for Native Forest Restoration

Where political, economic, social, and ecological considerations deem native forest restoration as the highest and best land use, existing radiata pine stands can play a central role in the transition. Modern forest restoration literature emphasizes the use of non-native or undesirable species to facilitate secondary succession (Ashton *et al.*, 1997; Kirkman *et al.*, 2007; Parrotta *et al.*, 1997). Plantation forests provide a "catalytic effect" in the understory microenvironment by modifying light quantity and quality, soil moisture, and nutrient availability. Careful manipulation and removal of the plantation overstory promotes the germination, establishment, and growth of advanced regeneration of desired native forest species (Ashton *et al.*, 1997; Parrotta *et al.*, 1997). Moreover, plantation forests can provide these ecosystem benefits while continuing to generate important economic benefits through plantation overstory removal harvests (Oberhauser, 1997; Parrotta *et al.*, 1997).

Understanding the functional relationship between the plantation forest and regeneration/succession processes is essential for creating appropriate silvicultural prescriptions within a restoration pathway. Research on species diversity in Monterey pine plantations along the coastal zone of the Basque Country indicates a significant positive correlation between plantation age and species richness (Onaindia and Mitxelena, 2009). Monterey pine plantations have greater mean species richness and higher diversity values than mixed oak and mixed beech forests of the same age. These results are attributed to several factors including i) the increased shade of mixed beech and mixed oak forests, ii) the increased presence of generalist and pioneer species in Monterey pine stands, and iii) the occurrence of characteristic mixed oak forest species in Monterey pine stands. In fact, the presence of many early-, mid-, and late-successional native tree species (e.g., *Betula*, *Castanea*, *Corylus*, *Crataegus*, *Fagus*, *Quercus*, *Rhamnus*, and *Salix*), as well as the late-successional fern species *Dryopteris affinis* Fraser-Jenkins, indicates the advanced successional development stage possible within older Monterey pine plantations.

These findings, as well as empirical evidence from Aiako Harria Nature Park, suggest that mature Monterey pine plantation forests do facilitate secondary succession of native forest by supporting significant communities of native forest species. Research from other forested regions supports this conclusion and highlights several additional factors that affect the rate of understory colonization, regeneration, and succession within plantations. These include i) plantation characteristics (e.g. age, spacing, structure, and overstory composition), ii) the presence of and proximity to native forest stands and seed sources, iii) reliable vectors of seed-dispersal (e.g. wind and birds), and iv) protection from chronic disturbance (Parrotta, 1995).

Plantation Characteristics

Research on native forest restoration within pine plantations in Thailand and South Africa indicates that plantation forests become more complex as they age, acquiring multiple canopy layers with increasing species richness, stem density, and tree sizes (Geldenhuys, 1997; Oberhauser, 1997). This corroborates results from research of Monterey pine plantations in New Zealand, which found that richness and cover of non-native, pioneer species decreases with plantation age, while richness and cover of native species increases. This is attributed to species' response to changes in the structure of the pine canopy that affect light quantity and quality in the understory (Brockerhoff *et al.*, 2003).

Research from Uganda comparing native forest restoration potential beneath pine and cypress plantations found higher levels of native species advanced regeneration below pines. Again, the difference in native species richness is attributed to the growth, development, and structure of the pine overstory. Pines grow faster and taller than cypress, forming a closed canopy earlier that shades out competing vegetation. As the plantation then ages, pine (unlike cypress) sheds its lower branches and develops taller clean boles, which provides the light regime and growing space necessary for understory re-initiation of native tree species (Fimbel and Fimbel, 1996; Geldenhuys, 1997).

Plantation spacing also influences the establishment, growth and survival of native tree species regeneration within plantations. Ashton *et al.* (1997; 1998) found that species planted in two- and three-rowwide gaps had the greatest height growth and dry masses, indicating that the modified light regime from canopy treatments is important for native species establishment and growth. Likewise, tree species which form a thin canopy of fine, readily decomposable leaf litter create a light- to moderately-shaded understory that suppresses grasses and other competing vegetation while allowing for the germination and growth of native species (Parrotta, 1995).

In the Basque Country where multiple conifer species are used in plantation forestry operations (e.g., *Chameacypris*, *Larix*, *Picea*, *Pinus*, and *Pseudotsuga*), these results suggest that species that self-prune, develop longer clean boles, and have less dense crowns are better facilitators of understory re-initiation and secondary succession. Restoration strategies must take into account the relationship between these factors and the autecological requirements of the desired native forest species. A range of silvicultural treatments that manipulate the pine overstory can create the various understory environments appropriate for the establishment, growth, and succession of a diverse mix of native species guilds.

Native Forest Remnants

The rate of understory colonization and natural regeneration within plantations is also influenced by the presence of and proximity to remnant native forest stands. Native forests are a crucial seed source for dispersal of native species. Therefore, the degree of fragmentation and isolation of native forest remnants affects dispersal potential and species richness within plantations (Onaindia and Mitxelena, 2009). If seed sources for important native species (especially fruit and nut bearing species that attract seed-dispersing wildlife) are not present in native forest fragments, then restoration may also require enrichment plantings of these species within the plantation (Ashton *et al.*, 2001; Kenk and Guehne, 2001). Hence, the location and arrangement of native forest remnants, as well as species diversity within those remnants, should be taken into account when deciding where forest restoration efforts will be prioritized across the Basque landscape.

Seed Dispersal

Research indicates two major mechanisms for seed dispersal into plantations: wind and wildlife. Small, light, and abundantly seeded species with wing-like attachments are adapted to wind transportation, allowing the species to colonize a wide area around the seed source (Parrotta, 1995; Geldenhuys, 1997). Examples from the native forests of Aiako Harria Nature Park include species like *Acer*, *Betula*, *Salix*, and *Ulmus*. Medium to large and less-abundantly seeded species often have nuts or are contained within fleshy fruits that attract wildlife, including rodents and birds, which then carry and disperse seed within plantations (Parrotta, 1995; Geldenhuys, 1997). Examples from Aiako Harria Nature Park include species like *Quercus*, *Castanea*, *Ilex*, *Prunus*, *Sorbus*, and *Taxus*.

Assuming a nearby seed source, wind-dispersed species will enter plantations and regenerate naturally where disturbance has scarified the understory and exposed mineral soil microsites for germination. Animal-dispersed seeds, on the other hand, rely on the movement of animals into and out of the plantation, which is a more stochastic dispersal vector. Fimbel and

Fimbel (1996) found, however, that the open understory and open canopy structure of pine plantations facilitate the movement of wide variety of wildlife species, thereby increasing rates of seed dispersal. These results suggest two ways to increase the rate of understory colonization and enhance natural regeneration in Basque plantation forests. First, use enrichment plantings to introduce both wind-dispersed species that are absent or too distant to colonize naturally and animal-dispersed species that will attract wildlife and promote their movement into and out of the plantation. Second, protect important wildlife species and their habitat in and around native forest remnants and restoration areas to increase the potential for this vector of seed transport (Oberhauser, 1997; Parrotta, 1997).

Chronic Disturbance

As mentioned previously, continuous, bottom-up and top-down impacts lead to chronic forest degradation, which disrupts understory re-initiation, succession, and stratification. Research from temperate forests in central Europe indicates that chronic disturbance, especially grazing, is a major inhibitor in the development of native broad-leaved forest species in the understory of coniferous plantations (Zerbe, 2002). This suggests that protecting potential restoration areas from livestock grazing is of paramount importance to any restoration strategy. In Aiako Harria Nature Park, exclosures around priority forest restoration areas in conjunction with the restoration of productive, enclosed pasturelands could reduce grazing pressure within native forest areas, promote natural regeneration, and reinstate natural succession processes.

Silvicultural Prescriptions for Native Forest Restoration

As mentioned above, effective silvicultural treatments imitate the natural disturbance regime, which underscores the importance of utilizing the appropriate treatment for a given site and forest type to restore the desired species composition. The number and distribution of trees removed, the method of removal (e.g., harvesting, girdling, and pull-downs), and the rate at which they are removed directly affects the understory microenvironment.

Deliberate and well-planned silvicultural treatments use these factors to drive understory re-initiation and successful regeneration of the desired and appropriate native forest floristic community. Based on empirical evidence from Aiako Harria Nature Park, there are two primary silvicultural systems that merit testing in current efforts to convert pine plantations to native forests.

The first system utilizes medium to large regeneration harvests followed by intensive, mixed-species reforestation, or what was referred to previously as full ecosystem restoration. This treatment is perhaps the simplest silvicultural prescription to quickly transform exotic plantation forests, and may be appropriate for regenerating early- and mid-successional, shade intolerant tree species on hotter, xeric sites characterized by frequent, large, and more severe disturbances. As previously mentioned, full ecosystem restoration of this nature is often prohibitively expensive and therefore not replicable at a forest or landscape scale. However, on sites in Aiako Harria Nature Park where the natural forest type is mixed oak (i.e., southern aspects, from mid- to upper-slope), this silvicultural prescription can be adapted to remove patches of overstory trees, in larger openings, and over a shorter time period. This will facilitate the understory microenvironment necessary regenerate a broad suite of shade intolerant, early- and mid-successional species (Figure 3) (Kenk and Guehne, 2001; Onaindia and Mitxelena, 2009).

The second system utilizes girdling, single-tree, and group selection harvests to create smaller canopy openings that imitate the small, non-lethal disturbances characteristic of cooler, mesic sites and regenerate a mix of mid- to late-successional, shade tolerant species. On sites in Aiako Harria Nature Park where the natural forest type is mixed beech (i.e. northern aspects from drainage to mid-slope), this silvicultural prescription can be used to gradually remove the pine overstory in order to promote an understory micro-environment that favors advance regeneration of shade-tolerant, mid- and late-successional species (Figure 3) (Kenk and Guehne, 2001; Onaindia and Mitxelena, 2009).

Historically, the native Basque forest landscape was a heterogeneous mosaic of forest types driven by a unique combination of biophysical site characteristics and dynamic forest processes. An appropriate restoration pathway accounts for this heterogeneity, drawing upon the principles of forest ecology and ecological silviculture to implement a range of silvicultural treatments in various spatial and temporal arrangements across the forested landscape. Utilizing such a holistic strategy will facilitate the regeneration of multiple age classes and guilds of native trees and promote the controlled conversion of existing plantation forests.

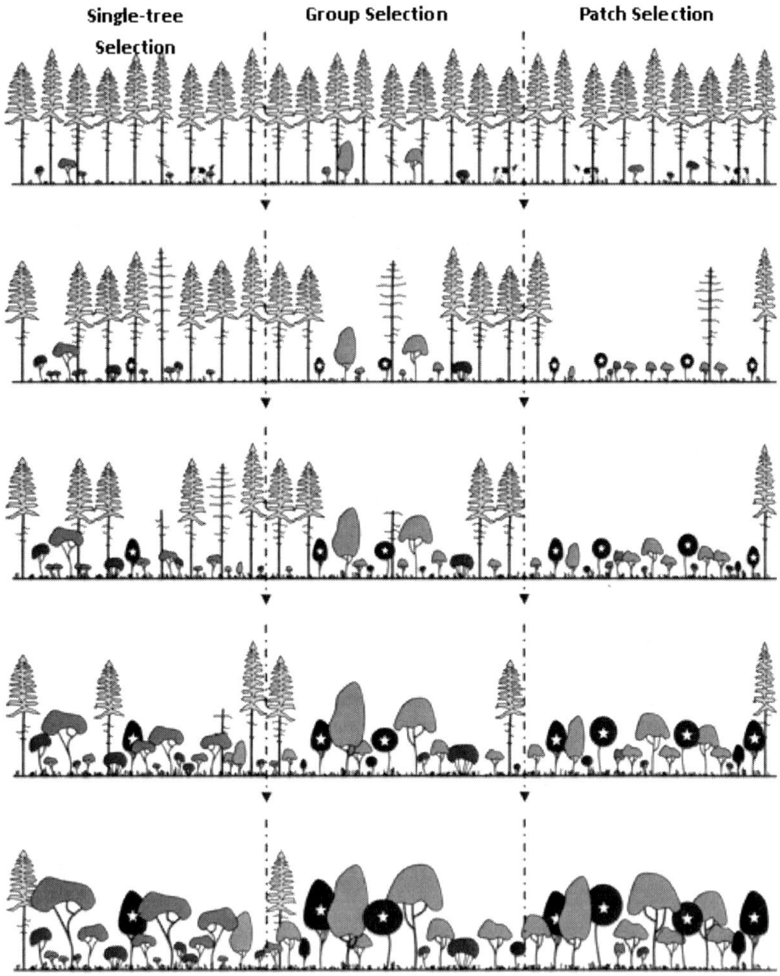

Figure 3. A range of silvicultural treatments intended to imitate the natural disturbance regime of a given site are necessary to convert pine plantation forests to complex, mixed-species deciduous forests in the Aiako Harria Natural Park. The size, arrangement, and rate of overstory removal allow for a controlled manipulation of the forest understory environment, facilitating the regeneration of the desired and appropriate suite of native forest species. Other key components of a successful restoration pathway include the control of chronic disturbance (i.e., removal of livestock) and enrichment planting (indicated by starred species above) to reintroduce important tree species.

This in turn will facilitate the re-establishment of complex forest structures, function, and resiliency in a restored secondary native, mixed-species deciduous forest. As Kenk and Guehne (2001) state, this "close-to-nature-silviculture" is intended to convert conifer plantations in a deliberate way following the tenet of establishing a "natural forest on its natural site."

REFERENCES

Allen H. L., Fox, T. R., Campbell, R. G. (2005). What is ahead for intensive pine plantation silviculture in the south?. *Southern Journal of Applied Forestry,* 29, 62-69.

Ashton, P. M. S., Gamage, S., Gunatilleke, I. A. U. N., Gunatilleke, C. V. S. (1997). Restoration of a Sri Lankan rainforest: using Caribbean pine *Pinus caribbaea* as a nurse for establishing late-successional tree species. *Journal of Applied Ecology,* 34, 915-925.

Ashton P. M. S., Gamage, S., Gunatilleke, I. A. U. N., Gunatilleke, C. V. S. (1998). Using Caribbean pine to establish a mixed plantation: testing effects of pine canopy removal on plantings of rain forest tree species. *Forest Ecology and Management,* 106, 211-222.

Ashton, M. S., Gunatilleke, C. V. S., Singhakumara, B. M. P., Gunatilleke, I. A. U. N. (2001). Restoration pathways for rain forest in southwest Sri Lanka: a review of concepts and models. *Forest Ecology and Management,* 154, 409-430.

Amezaga, I. and Onaindia, M. (1997). The effect of evergreen and deciduous coniferous plantations on the field layer and seed bank of native woodlands. *Ecography,* 20, 308-318.

Barten, P. K., Kyker-Snowman, T., Lyons, P. J., Mahlstedt, T., O'Connor, R., Spencer, B. A. (1998). Massachusetts: Managing a Watershed Protection Forest. *Journal of Forestry,* 96, 10-15.

Brockerhoff, E. G., Ecroyd, C. E., Leckie, A. C., Kimberley, M. O. (2003). Diversity and succession of adventive and indigenous vascular understorey plants in *Pinus radiata* plantation forests in New Zealand. *Forest Ecology and Management,* 185, 307-326.

Chauchard Badano, L. M. (2000). Growth and production of *Pinus radiata* D. Don plantations in the province of Guipuzco (Basque Country). Doctoral dissertation. Escuela Tecnica Superior de Ingenieros de Montes, Madrid, Spain.1-184.

Clements, F. E. (1916). Plant succession: An analysis of the development of vegetation. Publication No. 242, Carnegie Institution, Washington.

Euskadi Forestal. (2005). *Diagnosis and future of the Basque forest sector.* Retrieved November 6, 2009, from http://www.basoa.org/revista.asp.

Felix, A. C., Sharik, T. L., McGinnes, B. S., Johnson, W. C. (1983). Succession in loblolly pine plantations converted from second-growth forest in the central Piedmont of Virginia. *American Midland Naturalist,* 110, 365-380.

Fimbel, R. A. and Fimbel, C. C. (1996). The role of exotic conifer plantations in rehabilitating degraded tropical forest lands: A case study from the Kibale Forest in Uganda. *Forest Ecology and Management,* 81, 215-226.

Foster, D. R. (1992). Land-use history (1730-1990) and vegetation dynamics in central New England, USA. *Journal of Ecology,* 80, 753-772.

Foster, D. R. and Boose, E. R. (1992). Patterns of forest damage resulting from catastrophic wind in central New England, USA. *Journal of Ecology,* 80, 79-98.

Garcia, D., Quevedo, M., Ramon Obeso, J., Abajo, A. (2005). Fragmentation patterns and protection of montane forests in the Cantabrian range (NW Spain). *Forest Ecology and Management,* 208, 29-43.

Garcia Perez, J. D. and Groome, H. (2000). Spanish forestry planning dilemmas: technocracy and participation. *Journal of Rural Studies,* 16, 485-496.

Garcia-Ruiz, J. M., Lasanta, T., Ruiz-Flano, P., Ortigosa, L., White, S., Gonzalez, C., Marti, C. (1996). Land-use changes and sustainable development in mountain areas: a case study in the Spanish Pyrenees. *Landscape Ecology,* 11, 267-277.

Geldenhuys, C. J. (1997). Native forest regeneration in pine and eucalypt plantations in Northern Province, South Africa. *Forest Ecology and Management,* 99, 101-115.

Howard, L. F. and Lee, T. D. (2002). Upland old-field succession in southeastern New Hampshire. *Journal of the Torrey Botanical Society,* 129, 60-76.

Kenk, G. and Guehne, S. (2001). Management of transformation in central Europe. *Forest Ecology and Management,* 151, 107-119.

Kirkman, L. K., Mitchell, R. J., Kaeser, M. J., Pecot, S. D., Coffey, K. L. (2007). The perpetual forest: using undesireable species to bridge restoration. *Journal of Applied Ecology,* 44, 604-614.

Montiel, C. and Galiana, L. (2005). Forest policy and land planning policy in Spain: a regional approaChapter *Forest Policy and Economics,* 7, 131-142.

Mosandl, R. and Kleinert, A. (1998). Development of oaks (*Quercus petraea* (Matt.) Liebl.) emerged from bird-dispersed seeds under old-growth pine (*Pinus silvestris* L.) stands. *Forest Ecology and Management,* 106, 35-44.

Mosandl, R. and Kussner, A. (1999). Conversion of pure pine and spruce forests into mixed forests in eastern Germany: some aspects of silvicultural strategy. In A. F. M. Olsthoorn, H. H. Bartelink, J. J. Gardiner, H. Pretzsch, H. J. Hekhuis, A. Frank (Eds.), *Management of Mixed-Species Forests: Silviculture and Economics* (pp. 208-218). IBN Science Contribution, 15.

National Forest Inventory. (1996). *Second national forest inventory.* Retrieved

November 6, 2009 from http://www.mma.es/portal/secciones/biodiversidad/ inventarios/ifn/ifn2/imdex.htm.

Oberhauser, U. (1997). Secondary forest regeneration beneath pine (*Pinus kesiya*) plantations in the northern Thai highlands: a chronosequence study. *Forest Ecology and Management,* 99, 171-183.

Oliver, C. D. and Larson B. C. (1996). Forest Stand Dynamics. New York: Wiley.

Oliver, C. D. and Meznick, R. (2005). Investing in Forestry. *Journal of Forestry,* 21, 97-111.

Onaindia M. and Mitxelena, A. (2009). Potential use of pine plantations to restore native forests in a highly fragmented river basin. *Annals of Forest Science,* 66, 305.

Pardo F., Gil, L., Pardos, J. A. (2004). Structure and composition of pole-stage stands developed in an ancient wood pasture in central Spain. *Forestry,* 77, 67-74.

Parrotta, J. A. (1995). Influence of overstory composition on understory colonization by native species in plantations on a degraded tropical site. *Journal of Vegetation Science,* 6, 627-636.

Parrotta, J. A., Turnbull, J. W., Jones, N. (1997). Catalyzing native forest regeneration on degraded tropical lands. *Forest Ecology and Management,* 99, 1-7.

Raup, H. M. (1966). The view from John Sanderson's farm: A perspective fro the use of land. *Forest History,* 10, 2-11.

Rescia, A. J., Schmitz, M. F., Martin de Agar, P., De Pablo, C. L., Pineda, F. D. (1997). A fragmented landscape in northern Spain analyzed at different spatial scales: Implications for management. *Journal of Vegetation Science,* 8, 343-352.

Schmitz, M. F., J. A. Atauri, C. L. De Pablo, P. M. De Agar, A. J. Rescia, F. D. Pineda. (1998). Changes in land use in Northern Spain: Effects of forestry management on soil conservation. *Forest Ecology and Management,* 109, 137-150.

Scott, C. W. (1960). Radiata pine as an exotic. *Unasylva,* 14, 6-16.

Yin, R. and Sedjo, R. A. (2001). Is this the age of intensive management?: A study of loblolly pine on Georgia's Piedmont. *Journal of Forestry,* 99, 10-17.

Zerbe, S. (2002). Restoration of natural broad-leaved woodland in Central Europe on sites with coniferous forest plantations. *Forest Ecology and Management,* 167, 27-42.

In: Multiple-Use Management ... ISBN 978-1-61942-328-2
Eds: A. DuVal, T. Muruetagoiena et al.

Chapter 4

FINANCIAL AND ECOLOGICAL IMPLICATIONS OF TRANSITIONING *PINUS RADIATA* PLANTATIONS TO MIXED-SPECIES *FAGUS SYLVATICA* AND *QUERCUS SPP.* FORESTS IN THE BASQUE COUNTRY, SPAIN

Sarah A. Lowery*
ECODIT LLC
Washington, District of Columbia, US

ABSTRACT

The forests of the Basque Country, Spain, are composed primarily of single-species plantations with exotic conifers. These plantations have proved financially viable for private and public landowners, although a recent downturn in the Spanish timber markets and an increased focus on

---

* Sarah Lowery completed her Masters of Environmental Management and Masters of Business Administration degrees in 2010 at the Yale School of Forestry and Environmental Studies and the Yale School of Management, New Haven, CT USA. The author is incredibly grateful towards Aitor Onaindia for sharing relevant literary resources and answering many questions relating to the following analysis, as well as Jon Ugarte, Josean Belzunce, and Carlos Aseginolaza for sharing their technical expertise and experience in the management of Monterey pine plantations.

managing forests for a diverse set of values has led landowners to consider whether converting these plantations to native, mixed-species deciduous forests may be financially or ecologically attractive. This paper analyzes current market conditions in the Basque Country for timber, particularly Monterey pine, and assesses the financial implications of four possible management alternatives: (1) harvest now despite the low timber price and replant with pine; (2) harvest in five to ten years and then replant with pine; (3) harvest now and replant with beech and oak seedlings; or (4) encourage natural regeneration of beech and/or oak and harvest the pine overstory in five to ten years. Then follows a discussion of the financial risks and the ecological implications involved in cultivating Monterey pine as opposed to a mix of beech and oak. The analysis concludes that while cultivation of mixed hardwoods is not more financially attractive, the combined economic and ecological benefits of such forests may be attractive to landowners.

Keywords: Financial analysis, forest management, risk, timber, even-aged plantation forestry, uneven-aged mixed-species natural forests, transition

INTRODUCTION

The forests of the Basque Country, Spain, are predominantly even-aged coniferous plantations of Monterey pine (*Pinus radiata* D. Don), Douglas fir (*Pseudotsuga menziesii* (Mirb.) Franco), or larch (*Larix* L.), in addition to a few other exotics such as Eucalyptus (*Eucalyptus* spp.) and red oak (*Quercus rubra* L.) and the occasional stand of centuries-old pollarded European beech (*Fagus sylvatica* L.) The Spanish government promoted an afforestation policy in the 1940s and 1950s that led to large portions of public lands being planted with various tree species, including fast-growing pines such as Monterey pine (Scott, 1960). Additionally, private landowners cultivate non-natives: approximately 85.4 percent of Monterey pine plantations are privately owned (Inventario Forestal, 2005).

In more recent years, however, there has been a shift in some countries' policies away from planting exotic, single-species plantations. For instance, Germany has encouraged a transformation from pure to mixed-species stands (Hanewinkel, 2001). Many European countries now have forest management policies designed to promote the conversion of coniferous, single-species forests into deciduous or mixed-species forests (Zerbe, 2000; Knoke, Ammer, Stimm, and Mosandl, 2008). Although the Spanish forest policy has not been modified to follow suit, local policy in Aiako Harria Nature Park stipulates that only native species can be planted on public lands.

Many of the reasons for these changing forest policies have focused on ecological factors such as the water quality, increased biodiversity, resilience, and structural diversity of the resulting forests (Knoke et al., 2008). Economic considerations are also important for public landowners who are faced with real financial constraints, especially regarding transition costs, as they develop forest and land management practices. Furthermore, private landowners – to the extent that they are managing their land to generate the highest financial return possible – have even more of an incentive to take into account the financial implications of transforming single-species and single age-class stands into mixed-species and/or uneven-aged forests.

This paper attempts to outline the financial implications of maintaining Monterey pine plantations versus transitioning these forests into the type that were once common across the Basque Country landscape: mixed-species stands dominated by native species such as European beech (*Fagus sylvatica* L.) or Pendunculate oak (*Quercus robur* L.). The analysis begins by exploring current market conditions in the timber industry in the Basque Country and Southern Europe overall. Then a net present value analysis is conducted for several scenarios of either replanting Monterey pine or transitioning these plantations into native oak or beech stands. This includes a discussion of silviculture treatments for each scenario and an annotated review of the literature on the financial implications of single species, single age-class stands vis-à-vis mixed-species, multiple age-class stands. Given that most investors are risk-averse (Ross et al., 2008), a discussion of the various risks involved in each type of silviculture regime is undertaken. Finally, other values such as water quality, biodiversity, and carbon sequestration are considered in deciding whether to transition Monterey pine plantations to native, mixed-species forests.

The Timber Industry and Current Market Conditions

The timber industry in the Basque Country is fairly fragmented due to small plantation size and numerous owners. For instance, about 25,000 proprietors manage 150,000 hectares of Monterey pine. The average plantation size is six hectares, and 73 percent of plantations are actually comprised of less than five hectares (Schmitz et al., 1998; Chauchard, 2000). There appear to be few barriers to entry, as anyone with a little land – even less than a hectare – can plant seedlings and acquire basic knowledge of cultivating timber through resources like the Confederation of Foresters in the Basque Country (the organization's website, www.basoa.org, has a plenitude of information on silvicultural treatments for Monterey pine and other species). Many owners have traditionally maintained their plantations themselves, including site preparation, thinning work, and harvesting the timber. However, in more recent years, the trend has been to hire foresters to do much of this work. Unlike older generations of farmers who worked the land and maintained the plantations, the majority of these landowners have less forestry knowledge and also hold jobs or own other businesses that require their time. This, combined with the existence of many small plantations, means that plantation owners have virtually no supplier power in the market and are thus price takers when it comes time to harvest and sell their timber.

In contrast, there are relatively few forestry companies in the Basque Country that are available to complete the work of site preparation, thinning, clearing, pruning, and harvesting. For instance, the biannual report of the Confederation of Foresters in the Basque Country – Euskadi Forestal (September, 2009) – lists about 40 such firms. Furthermore, these companies tend to specialize in a portion of the silvicultural work available – e.g., clearings or the final harvest of sawtimber – and are territorial in nature; in fact, disputes often arise between these small- or medium-sized enterprises over the areas in which they work. Anecdotal evidence from the Basque Country suggests that foresters are capitalizing on their relative advantage in the market (vis-à-vis plantation owners) by colluding and offering low prices for the timber they harvest from these plantations. The only recourse for the owners is to harvest the timber themselves or wait to do so until supplier/buyer conditions change in their favor, although neither option is attractive or necessarily viable.

Current market conditions in the timber industry are quite poor; in fact, landowners are purportedly not harvesting their mature Monterey pine plantations because prices are so low. Prices of Monterey pine for thick sawtimber fell 30 percent between January and September of 2009 to just under 32€ per m^3 (Euskadi Forestal, September, 2009). Various factors likely contributed to the current depressed prices. First, Hurricane Klaus destroyed more than 40 million m^3 of timber in the Landes region in nearby France in January of 2009, and Euskadi Forestal (September, 2009) – as well as reports from the Basque Country – attribute the fall in prices primarily to the overabundance of salvaged timber flooding the market after the hurricane. (However, the journal also states that the real price of wood in 2007, before either Hurricane Klaus or the economic crisis, had already fallen to less than half the level in 2001, indicating that other factors have also contributed to the fall in prices.) Secondly, according to a report from Global Wood on the European wood product industry, causes for a weak timber market in Spain include a sudden halt in the country's construction boom in 2008 that has left Spain with "seven times more housing than the current need" (Wood Product Prices, 2009). The impact of a weak construction market on timber prices in Spain is probably less dramatic than in the United States or the United Kingdom, where wooden construction is more common, but it nevertheless probably contributes to lower prices. Third, anecdotal evidence from the Basque Country indicates that inexpensive timber – including Monterey pine – imported from countries such as Siberia, New Zealand, and Paraguay also depresses local prices in the Basque Country.

The market for oak and beech is relatively undeveloped in the Basque Country provinces of Gipuzkoa, Bizkaia, and Àlava. Since rotation lengths for these deciduous species is much longer than that of Monterey and other conifers, this may have been a deterrent to planting these trees during the mid-century reforestation surge. The markets for these species are more developed in France, however, and prices for oak and beech in France are used as proxies for the prices that plantation owners in the provinces mentioned above might expect should they cultivate these species. Figure 1 shows historic prices for all three species in real terms, and Table 1 details the mean, standard deviation, and trend line for each species. As you can see, the real prices of Monterey pine, oak, and beech have fluctuated over the years like most commodities, although the overall trend has been positive. The mean price for oak is more than double either of the other two species, although its standard deviation – a measure of spread and variability – is correspondingly greater.

Depressed market conditions for Monterey pine may continue for several years, but evidence from studies such as Prestemon and Holmes (2000) suggests that low prices caused by an influx in timber due to a disaster like Hurricane Klaus are most likely a short-term effect and the long-term implications may actually lead to enhanced prices. Of course, other variables should also be considered when estimating future prices, such as increases in cheap timber supplies from Eastern Europe, which may endure for a longer time period.

Source: Revue forestière française, 1920-2009; Euskadi Forestal, 1984-2009.

Figure 1. Timber prices, by Species.

Table 1. Historic Monterey pine, oak, and beech prices (Constant 2005 €/m3)

Species	Mean	Standard Deviation	Slope
Monterey pine	65	13.3	.033
Oak	158	57.1	.683
Beech	77	23.7	.379

Financial Analysis

The financial value of managing land for timber production – like other potential investments – can be estimated by calculating the net present value (NPV) of the expected future cash flows generated from forest stands (Equation 1).

$$NPV = \sum_{y=0}^{n} \frac{R_y}{(1+r)^y} - \sum_{y=0}^{n} \frac{C_y}{(1+r)^y} \tag{1}$$

here, R_y equals the revenues associated with thinning and final harvests, C_y incorporates all the costs of cultivating the trees (such as planting, site preparation, weed control, and thinning), and r is the discount rate.

If the future value of the land itself is also taken into account in the NPV analysis by assuming land sale after a certain number of rotations, this value is known as the willingness to pay for land or WPL (Klemperer, 1996). Although private landowners in particular would want to take into account the value of the land itself when making management decisions, this analysis focuses on the financial value of the two different stand types rather than introducing the third variable of land value into the equation for simplicity of analysis and to focus on what is often the main determinant of the difference in financial value between the two stands – the value of the timber harvest. Also, anecdotal evidence from the Basque Country suggests that although landowners may allow land to become fallow should agricultural or timber production prove financially unviable, land sale is rare in this instance: landowners will simply wait until cultivation of crops or timber becomes lucrative again. Thus, it appears that landowners are relatively unlikely to take the possible sale of the land after the harvest into account in their management decisions. This study is geared towards these current landowners and the decisions they face, including public land managers who rarely have the option to sell the land. Thus, an NPV analysis, such as the one presented here, more accurately presents these public land managers' choices than a WPL analysis might.

Management Alternatives

In the present study, the economic implications of maintaining Monterey pine stands versus converting the forests into native beech or oak stands are estimated. The former option may be the easiest from a silvicultural perspective – i.e. the landowner already knows the schedule of treatments and thinnings for Monterey pine – and the latter transformation may be the most ecologically desirable. It should be noted that these are not the only options, however. For instance, private landowners may decide to cultivate other conifers like Douglas fir, which is less susceptible than Monterey pine to diseases like pitch canker (discussed later in this study). Alternatively, they may plant exotic hardwoods such as red oak or tulip poplar (*Liriodendron tulipifera* L.), or native species that never dominated Basque forests but could be financially profitable like wild cherry (*Prunus avium* L.), ash (*Fraxinus excelsior* L.), or walnut *(Juglans regia L.)*. Although an exhaustive financial analysis of these various alternatives is not possible in the current study, it would be enlightening to analyze these alternatives as well; perhaps a future study will undertake this investigation.

In addition, it should also be noted that the management alternatives presented in this paper apply primarily to those landowners who currently have mature Monterey pine plantations (ages 30-35 years). Of course it is impossible to know exactly how many landowners fall into this category, although it is not unreasonable to assume that many could be ready to harvest their second crops in the near future – if many planted their first Monterey pine plantations in the 1940s and 1950s on 30-35 year rotations. Moreover, foresters in the Basque Country indicate that many landowners do indeed have stands ready to be harvested. As Nichols et al. (2006) points out, reliable financial data on timber plantations are difficult to obtain and few publications delve into economic comparisons of the two types of plantations. Cost and revenue data gathered on site in the Basque Country, as well as supplemental data from journal articles and industry research (detailed below) informs the assumptions in the economic evaluation. As detailed above, this study looks at four management alternatives that plantation owners in the Basque Country may have before them:

1. Harvest now and replant with Monterey pine;
2. Wait several years until the price potentially rises for Monterey pine and then replant;
3. Harvest now and replant with beech or oak seedlings; or
4. Cultivate natural regeneration of oak and/or beech beneath Monterey plantations, harvest the pine when the price has risen, and subsequently harvest the oak or beech when it is mature.

The first two options will maintain the plantations as Monterey pine, and the second two will allow the conversion to mixed stands dominated by native species of oak and/or beech. After a discussion of the prescribed silviculture regime for these two sets of options, each alternative is considered.

Maintaining Monterey Pine Plantations

Silviculture Prescriptions

Monterey pine has been cultivated as an exotic species for almost a century in countries around the world such as Chile, New Zealand, Australia, South Africa, and Spain (Scott, 1960). Due to this prodigious experience managing the species, it is not surprising that most cultivators follow a similar set of treatments for even-aged Monterey pine stands, including those in the

Basque Country. A classical treatment cycle for Monterey pine includes a series of steps: first, the site is prepared through methods such as burning slash, scalping, or scalping and down-slope ripping (Olarieta et al., 1999) and seedlings are planted in 2-2.5 by 4 m^2 plots; second, a thinning of stems is undertaken in years 9-13; third, a second thinning occurs in years 17-19 (with a third thinning occurring in years 23-27 if the rotation length is a bit long); and fourth, the stand is harvested (Otazua, 1998; Hanewinkel, 2001; anecdotal evidence from the Basque country; SECOMA). More intensive management also stipulates low and high pruning in years 9 and 13 (Pino Insignis Fichero, *n.d.*) in an attempt to produce timber without the knots created if branches are left growing on the tree. The clear timber produced if pruning is undertaken increases the value of the timber but also increases management costs. Some sources in the Basque Country say that pruning is usually undertaken, although the majority reports that pruning rarely occurs.

One would imagine that financial evaluation of Monterey pine plantations is fairly straightforward due to this standard cycle of treatments. However, estimates of costs and revenues vary according to the source (Otazua, 1998; various Basque Country sources), as do the years in which treatments are performed and the intensity of management (i.e. how many treatments are undertaken). The purpose of this study is to compare the differences in financial returns between Monterey pine and mixed-species, beech and oak stands rather than various management regimes for pine; therefore, costs and volume estimates across sources are averaged, as are the years in which treatments are undertaken (e.g. if the first thinning could be undertaken in years 9 through 13, year 11 is used in the analysis). Although pruning and cleaning may increase the value of the timber quite substantially and thus the profitability of the investment (Agirre, 2009), it appears that low intensive management regimes with few cleanings and pruning are more likely to be undertaken in the Basque region of Spain and are thus reflected in the analysis. A standard 35-year rotation length is assumed, although shorter or longer rotations may be chosen depending on various factors such as timber prices, site quality, etc. Yearly management costs (not including specific treatments such cleaning, pruning, and thinning) in the Basque Country are approximately 6€ per hectare for any management regime regardless of the species cultivated and are thus not included in the financial analysis.

An additional complication when estimating the financial costs or revenue associated with cultivating timber is the presence of subsidies in the three provinces of Àlava, Gipuzkoa, and Bizkaia. The amounts of these subsidies vary across the three areas but – on average – they amount to about 35 percent of the costs of cultivating pine (e.g., site preparation, cleaning, and pruning costs). Thus, this analysis includes a 35 percent subsidy for replanting pine this year (i.e., costs of site prep and planting are discounted by 35 percent only for Alternative One), but it does not account for other subsidies going forward. Also, sources in the Basque Country report that the provincial government of Gipuzkoa recently decided to set aside 1.3€ million to subsidize the harvest of "overgrown" pine plantations that would command little or no value at the sawmill. However, this subsidy is not included in the analysis since the amount potentially given to each landowner is unknown. In addition, anecdotal evidence from the Basque Country suggests that the amount is not significant enough to induce many landowners to harvest their plantations. In general, forest subsidies may come in various forms and change from year to year, which make them difficult to plan for or include in financial analyses – especially as they relate to the harvest, which will not occur for many years (further discussion of subsidies is included in the section, Risk Consideration).

The choice of an appropriate discount rate is guided by the Capital Asset Pricing Model (CAPM – shown in Equation 2), which is often used to estimate asset or project-specific discount rates. "CAPM remains the standard for risk and return modeling" (Mendell, 2006). Instead of likening the risk of timber investments to the risk-free rate like Agirre (2009), the CAPM allows the rate to incorporate the riskiness of the timber asset by including the beta of Monterey pine prices vis-à-vis the stock market.

$$Re = Rf + ße\ (Rm - Rf) \quad (2)$$

here, R_e is the return on equity, or the discount rate. R_f is the risk-free rate, which is estimated to be the interest rate of 30-year Spanish government bonds (4.94 percent, as used in Agirre, 2009). R_m is the average return one could achieve by investing in the stock market, which is estimated at 6.8 percent for the period 1988-2009 for the MSCI World Index (MSCI World Index, 2010). The riskiness of the asset is captured in its beta ($ß_e$), which is the standardized covariance of the return of Monterey pine prices with the return on the market portfolio (here, the MSCI World Index). This $ß_e$, which is estimated to be 0.085, is then multiplied by the return of the market above the risk-free rate ($R_m - R_f$). Substituting values into the equation yields a discount rate of 5.10 for Monterey pine:

$$0.0510 = 0.0494 + 0.085\ (0.068 - 0.0494)$$

This is the discount rate used in the financial analysis of Monterey pine below (i.e., Alternatives One and Two). (See Ross et al. (2008) for detailed information about the CAPM, beta, risk premium, and related financial topics).

Alternative 1–Harvest Now and Replant with Pine

As stated above, the price of Monterey pine in the Basque Country is severely depressed because of factors such as the excess supply of timber across the border in France.

Table 2. Alternative One

Year	Treatment	Stems/ha	m3/ha	Price(€)	Costs (€)	Revenue(€)	Discounted
0	Harvest		383.50	14.12	-	5,416.67	5,416.67
	Site prep and planting costs	1250-1600			(1,254.50)	-	(1,254.50)
1	Cleaning				(437.50)	-	(416.27)
2	Cleaning				(437.50)	-	(396.07)
3	Cleaning				(437.50)	-	(376.85)
9	Low pruning				(400.00)	-	(255.64)
11	Thinning	600	20.25	47.93	(406.99)	970.58	326.09
18	Thinning	450	46.50	47.93	(934.56)	2,228.75	528.63
25	Thinning	325	89.25	47.93	(1,793.75)	4,277.75	716.28
35	Harvest		383.50	59.82	-	22,940.97	4,022.69
NPV							8,311.02

Note: *Site prep and planting costs include the site prep described previously, purchase of seedlings, digging holes and planting the site. Since most Monterey pine cultivators do not fence their properties, fencing costs are not included. **Costs of thinning are estimated per m3 and averaged: some estimate that thinning costs are 23.5€ (about 34 m3 cut for 800€), while others estimate costs at 16.7€ (15 m3 for 250€). Thus, the average cost of 20€ per m3 was multiplied by the amount removed to estimate the costs of each thinning. ***Prices are based on Bloc de Notas (2009), assuming that none of the stems thinned are greater than 28 cm and over 75% of the stems in the final harvest are greater than 28 cm (according to IKT SA, 2002).

In the September issue of Euskadi Forestal (2009), prices of sawtimber ranged from 30 to 45€ per ton, although sources in the Basque Country quote current maximum prices of 20€ per ton at the mill. Using the latter price, which equals about 14€ per m3, harvesting this year (in 2010) yields an NPV of almost 8,311€ (see Table 2). The NPV for this option is positive, although it is easy to see why plantation owners are not harvesting their Monterey pine plantations due to the extremely low timber prices.

Alternative 2–Wait 5 or 10 Years to Harvest

If owners wait several years until timber prices rebound to the level they were before the hurricane – e.g., to about 55€ per m^3 – the NPV of the investment will be greater. For instance, waiting 5 years yields an NPV of 17,832€ and harvesting after 10 years still gives the owners an NPV of 13,906€. Although a price rebound is never certain to occur, negative price spikes following hurricanes have led to long-run enhancement effects of 10 to 30 percent increases in prices (Prestemon and Holmes, 2000). Therefore, expecting prices to rebound to the price level before the hurricane is not unreasonable. Some practitioners postulate that leaving trees standing will decrease the quality of the trees and lead to lower prices in 5 or 10 years, and this is may indeed be the argument that a forest technician uses to persuade a landowner to harvest timber now, despite the low prices. However, a knowledgeable Basque Country forester has stated that the timber quality does not decrease appreciably before 45 years of age, and thus most plantations owners would do well to delay harvesting until timber prices rebound.

Converting Coniferous Plantations to Beech and Oak

Financial Comparisons in the Literature

The benefits of mixed-species stands have been noted in the literature for decades; for example, Toumey and Korstian (1937, p. 287 and 282) postulate, “a mixed stand is more easily transformed or modified to meet present or probable future demands of the market.” Numerous studies have been undertaken comparing the financial implications of even-aged versus uneven-aged and single species vis-à-vis mixed-species silvicultural systems, with varied results (see Table 3 for an annotated list of articles on this subject). Two general conclusions can be made. First, uneven-aged silviculture practices are usually more financially attractive because harvests are more frequent and often income is generated sooner during the time span evaluated. Since income is discounted to account for the time value of money, a Euro today is worth more than a Euro tomorrow or 30 years from now. One caveat is that costs may be greater and/or revenue may be less (in nominal Euros) for more frequent, smaller-volume, smaller-diameter harvests; this is particularly relevant in the Basque Country, which has many small stands of just a few hectares. It appears unlikely that the costs would negate the benefits of this management regime, but it is an important point to keep in mind.

Table 3. Studies on the financial implications of various management regimes

No	Author	Year	Location	Species	Management regime	Financial effect	Possible explanations
1	Haight	1985	Arizona	Ponderosa pine	Even vs. uneven-aged	+	Sooner, more frequent harvests
2	Michie	1985	North-central United States	Mixed hardwoods	Even vs. uneven-aged	+	Sooner, more frequent harvests
3	Niese and Strong	1992	Northern Wisconsin	Sugar maple (majority), other hardwoods	Even vs. uneven-aged	+	High residual volume and low percentage of below-grade volume in the residual
4	Knoke	1999	Kreuzberg Forest, Germany	Norway spruce, fir, beech	Even vs. uneven-aged	+	Higher quality wood and less firewood
5	Mohr and Schori	1999	Berne, Switzerland	Silver fir	Even vs. uneven-aged	+	Lower costs of uneven-aged management
6	Tarp, Helles, Holden-Anderson, Larsen, and Strange	2000	Denmark	Beech	Even vs. uneven-aged	+	Higher quality of logs in uneven-aged management regime
7	Valkonen and Valsta	2001	Finland	Norway spruce, birch	Single (spruce) vs mixed species	+	Shorter rotation length for birch
8	Hanne-winkel	2002	Black Forest, Germany	Silver fir, Norway spruce	Even-aged, pure spruce vs. uneven-aged, mixed-species	Mixed	Depends on site conditions and silvicultural treatments
9	Knoke, Stimm, Ammer, and Moog	2005	Germany	Norway spruce, European beech	Single (spruce) vs. mixed species	-	Longer rotation length for beech
10	Nichols, Bristow, and Vanclay	2006	Queensland, Australia	*Pinus* hybrid	Single vs. mixed-species	Uncertain	Higher costs with mixed-species could be overcome by a yield stimulus

Second, mixed-species plantations can be more economically advantageous under certain conditions. For example, if a longer-rotation species is intermixed with a shorter-rotation species, the more frequent income boosts of the latter will have a positive impact on the financial analysis. Even if mixing the two species causes higher mortality (see Valkonen and Valsta, 2001), the income generated earlier in the time period may offset this substantially. Also, certain tree species grow better when inter-planted with other species. For example, Forrester et al. (2005) and Kelty (2006) both find that productivity in mixed-species plantations may increase up to two-fold. Although a financial analysis of this increase in productivity was not undertaken in these studies, the result would be positive as long as the inter-planted tree's economic value was comparable to (or even a bit less than) the originally planted tree.

Silviculture Prescriptions

It appears that cultivation of beech and oak is not frequently undertaken in the Basque Country provinces of Gipuzkoa, Bizkaia, or Àlava (apart from ancient pollarded beech forests used for charcoal and firewood), especially when compared to the abundance of Monterey pine and other exotic softwood plantations. Some oak and beech stands can be found in neighboring Navarre, though, and the Basque Confederation of Foresters produces guidelines for growing these species (see Fagus Sylvatica Fichero, *n.d.*, and Quercus Robur Fichero, *n.d.*). Beech and oak are rarely planted together (intermixed), as oak tends to be thinned out by beech (personal communication, Basque Country forester). However, beech and oak may be planted in different orientations or slopes within the same stand or in adjacent stands, such as is reflected in this analysis.

Similar to Monterey pine but with much longer rotation lengths, the silvicultural treatments for these species include cleanings during the first 6 or 7 years, followed by low pruning in years 12 to 15, and subsequent thinning every 10-25 years afterwards. According to the articles cited above, beech is harvested when it reaches 120 years of age, whereas oak is left to grow until it reaches 150 years of age. Although volume growth is quite dependent on site characteristics such as soil quality, water quantity, etc., and therefore a specific site index would be ideal for accurately predicting the thinning and harvest yields for a mixed oak/beech forest, such an index proved elusive. Therefore, production tables from Madrigal et al. (1989) for beech in nearby Navarre are used in conjunction with the silviculture regimes outlined by the Basque Confederation of Foresters to estimate the volume removed at each thinning and harvest for the two species (the growth rate of oak in the Basque Country is about 75 percent that of beech).

Due to the extremely long rotation age of these species, it is not easy to determine the most appropriate discount rate to use. For example, there are no 100-year government bonds. In fact, for a landowner to decide to plant a species that will not be harvested for over one hundred years, she/he must have a very low discount rate indeed (she/he will not even be alive when the trees are harvested). Therefore, instead of using the CAPM to determine the discount rate, the two percent rate that is applied by Beltratti et al. (1996) for very long investment horizons is used.

Alternative 3–Harvest Pine Now and Plant Beech and Oak

In order to assess the economic implications of harvesting Monterey pine plantations this year and planting a mixed, native species forest dominated by beech and oak (see Caligiuri, in this issue, for a more detailed description of this and other native forest types in the Basque country), a financial analysis is presented in Table 4.

As the analysis shows, the NPV of Alternative Three is just over fifty percent of the value of Alternative One (harvesting pine now and replanting with pine), despite the lower discount rate. To compare the two alternatives across similar time frames, Alternative One could be repeated roughly four times to make the time spans similar and to apply the same discount rate. However, given that Alternative One has already proven more financially viable, this analysis is not necessary to draw the conclusion that Alternative Three is not viable compared to Alternative One from a purely financial perspective. Additionally, a landowner is really deciding between two alternatives with different time spans; i.e. if she plants Monterey pine now, in 35 years she will again decide in which species to invest and is therefore not committing to four rotations of Monterey pine if she chooses to replant pine now. Furthermore, should the analysis be undertaken without the assumption that Monterey pine will first be harvested from the site before oak/beech are planted (e.g., if the current land use is agricultural in nature), it will yield a negative NPV due to the lack of revenue from a timber harvest this year, which no investor should consider undertaking. On the other hand, if harvesting of pine is delayed five or ten years (similar to Alternative Two), the NPV of the project increases substantially; for instance, a five year delay yields an NPV of almost 18,000€.

Table 4. Alternative Three

				Large Stems		Small Stems				
Year	Species	Treatment	Stems/ha	m3/ha	Price (€)	m3/ha	Price (€)	Costs (€)	Revenue (€)	Discounted
0	Radiata	Harvest		383.5	14.1			-	5,417	5,417
	Beech/Oak	Site prep and planting costs	1355					(2,833)	-	(2,833)
1	Beech/Oak	Cleaning						(438)	-	(429)
2	Beech/Oak	Cleaning						(438)	-	(421)
3	Beech/Oak	Cleaning						(438)	-	(412)
4	Beech/Oak	Cleaning						(438)	-	(404)
5	Beech/Oak	Cleaning						(438)	-	(396)
6	Oak	Cleaning						(219)	-	(194)
12	Beech	Low pruning						(200)	-	(158)
15	Oak	Low pruning						(200)	-	(149)
25	Beech/Oak	Thinning	975					(450)	-	(274)
40	Beech	Thinning	400	0.3	19.0	3.0	12.7	(65)	43	(10)
45	Oak	Thinning	250	0.7	54.7	4.0	13.4	(93)	90	(1)
55	Beech	Thinning	275	2.0	19.0	7.7	12.7	(194)	135	(20)
60	Oak	Thinning	175	1.6	54.7	6.3	13.4	(158)	170	4
70	Beech	Thinning	200	6.1	28.7	6.2	12.7	(245)	252	2
75	Oak	Thinning	125	6.1	111.1	3.6	13.4	(194)	723	120
85	Beech	Thinning	125	9.2	38.4	4.3	12.7	(270)	406	25

				Large Stems		Small Stems				
Year	Species	Treatment	Stems/ha	m3/ha	Price (€)	m3/ha	Price (€)	Costs (€)	Revenue (€)	Discounted
100	Beech	Thinning	75	10.7	38.4	4.3	12.7	(301)	466	23
	Oak	Thinning	75	12.8	167.4	4.1	13.4	(340)	2,202	257
120	Beech	Harvest	150	306.9	38.4	37.9	12.7	-	12,266	1,139
150	Oak	Harvest	150	331.0	167.4	40.9	13.4	-	55,953	2,869
NPV										4,155

Notes: *Fifty percent of the seedlings planted are oak and fifty percent are beech **Site prep and planting costs are assumed to be the same as Monterey pine with two exceptions: first, site prep costs are likely higher when preparing to grow a different species, such as beech or oak, and this is included in the site prep costs; second, average fencing costs per hectare are included since deciduous species are at greater risk from ungulate grazing than Monterey pine (Zerbe, 2002) and should thus be fenced. ***Prices are based on Revue Forestière Française (2009) assuming that the smaller stems thinned or harvested are less than 25 cm in diameter. The larger stems are assumed to less than 40 cm (for beech) and 50 cm (for oak) up until 70 years of age; from 70-85, half of the larger stems harvested are assumed to be below these diameters and half are assumed to be above; and from age 85 onwards, it is assumed that all of the larger stems are greater than these diameters (based on observations in the Basque Country).

It is important to remember that this increase is based wholly on an expected increase in the price of pine in five years and does not reflect a financial gain within the silviculture regime of planting, thinning, and harvesting beech and oak.

Alternative 4–Natural Regeneration

Given that Alternative Three is substantially less financially attractive than Alternatives One and Two, the final alternative in our analysis – Alternative Four – considers the option outlined in Caligiuri (this issue) of encouraging natural regeneration of beech and oak under pine plantations and then harvesting the pine when prices have risen. A conservative assumption is that prices will have risen to their early 2009 levels ten years from now and that site prep and planting costs for beech and oak will therefore be approximately half of their former levels due to natural regeneration (except for fencing costs, which will be incurred this year to exclude grazing and create conditions for natural regeneration). Under these assumptions, the NPV of this alternative is about 17,650€. In addition, should this strategy be pursued even without the large financial boost from the Monterey pine harvest (e.g., under a scenario in which the Monterey pine or other coniferous trees are already 45 years or older and will be unable to command a high price ten years in the future), a positive NPV of 431€ is attainable. This is particularly helpful information for public landowners, who likely have greater incentives to manage forests for a diverse set of values than do private landowners.

RISK CONSIDERATION

As shown, the net present value of Monterey pine plantations is greater than mixed-species stands of beech and oak. However, various sources of uncertainty – such as potential natural disturbances, diseases, fluctuating timber prices, and government policies – which may have a substantial effect on the expected financial return of managing timberland have not yet been wholly considered in this analysis. Knoke et al. (2001) points out that many of the comparative economic studies on silvicultural treatments in Central Europe ignore these associated risks, and Valsta (1998) argues that unequal risks between different silvicultural management options should be taken into account in financial analyses. Since these risks are real constraints in forest management decisions, they are discussed in some detail here. First, an explanation of portfolio theory is presented, as well as a subsequent explanation of its applicability to timber investments. Then a discussion as to how these results may be applicable to forest management in the Basque Country is undertaken, with special attention paid to natural hazards and government forest policies.

Portfolio Theory Applied to Forestry Investments

Markowitz's seminal 1952 paper on the topic of portfolio theory, "Portfolio Selection," introduced this important economic theory and has been expanded upon by others since then, including Nobel Laureate William Sharpe (1964). The basic premise of the theory is that investing in more than one asset decreases the riskiness of the portfolio without decreasing its profitability *if the riskiness of the individual assets is not directly correlated.* As a simple, fictional example, consider a financial investment in stocks of winter coats and beach towels. Sales of winter coats are likely to increase during the winter; in contrast, beach towel sales will swell in the summer. It is clear that the sales of each of these products are negatively correlated, and therefore, it reasonable to assume that the financial return on these investments is also negatively correlated and holding both stocks reduces the portfolio's unsystematic risk.[1] Even financial investments in assets that do not have a perfectly negative correlation can reduce the portfolio's risk. For instance, there is no reason to assume that an investment in stocks of laundry detergent would be correlated with the investment in stocks of winter coats; in this case, the correlation between the investments would be zero. Furthermore, even assets with positive correlations will lead to a reduction in risk as long as the correlation is less than perfect (Ross et al., 2008).

This diversification effect can be applied to any investment portfolio that includes assets with varying degrees of risk correlation, including forestry investments. Zinkhan et al. (1992) discusses portfolio theory and management as it applies to forestry, and various studies evaluate timber risk and return vis-à-vis other financial asset classes (Thomson, 1991; Caulfield, 1998). Wagner and Rideout (1991) find that the majority of forest assets' variability is nonsystematic and diversifiable and thus would increase a well-diversified portfolio's expected return without substantially increasing the portfolio's

[1] Unsystematic, or diversifiable, risk is the risk that can be mitigated by diversifying the portfolio. In contrast, systematic risk, which is the risk inherent to the entire market, cannot be avoided through diversification (Ross et al., 2008, p. 299).

variance. One of the most important sources of risk in forestry is the variation of future stumpage prices, since revenue streams from forestry investments are determined primarily by the price of timber and volume harvested. While timber prices may be correlated from year to year (Knoke et al., 2001), various studies have shown great volatility in timber prices during longer time periods (e.g., Norstrom, 1975; Brazee and Mendelsohn, 1988; Haight, 1990; Washburn and Binkley, 1990; Lohmander, 2000). Such studies are very useful in framing our understanding of how timber price risk – which is a real and measurable risk as described above – can affect financial valuation of different stands.

Forestry Investment Diversifies Portfolio

In the case of the small forest owner (recall that 73% of plantations are under five hectares in the Basque Country), who is likely to also have another business or be elsewhere employed, the benefits from diversification can be seen in her/his investment portfolio and/or at the stand level. In the former case, investing in various ventures – including but not limited to her/his plantation – will decrease the risk profile of the entire investment portfolio if the timberland investment is not perfectly correlated with the other types of investments. For instance, timber prices may be correlated with investments such as the stock market to varying degrees. As we saw in the financial analysis for Monterey pine plantations, the return of Monterey pine timber prices vis-à-vis the return of the MSCI World Index was 0.085, which indicates that Monterey timber prices are almost not at all correlated with the stock market (perfectly correlated, uncorrelated, and perfectly negatively correlated assets have betas of 1, 0, and -1, respectively). This indicates that investing in (thus planting) Monterey pine is likely to decrease the risk of the forest owner's investment *based on this correlation coefficient*. In this way, the riskiness of planting Monterey pine (at least to the extent that timber prices determine its riskiness) has already been incorporated into the financial analysis.

However, other risks – or sources of risk reduction – have not been incorporated into the financial analysis. For instance, a landowner may lower the potential risk of a plantation by cultivating two tree species instead of one. Just as investing in different assets may lower the risk of an investment portfolio, planting tree species with different characteristics is a type of diversification. For instance, different species are more or less susceptible to natural hazards like windfall or insect attack. Knoke et al. (2008) finds that native species like beech and oak are far less susceptible to windthrow than exotic species like Spruce. Omsbry (2004) points out that Monterey pine in particular suffers from infestations of pitch cancer in contrast to many other trees, and pines like Monterey pine are more susceptible to fire as well. Tree species often have different rotation lengths, and timber prices at the time of harvest are thus uncorrelated. Even if both species are harvested in the same year, however, they may command different stumpage prices that may or may not be correlated to some degree. The least risk would be achieved by selecting two trees with negatively correlated stumpage prices. Additionally, uneven-aged plantations provide more regular, albeit smaller, revenue flows to owners that enable the overall NPV of projects to be less susceptible to uncertainties regarding timber prices.

From simply looking at the graph of timber prices across the species studied (Figure 1), one can see that timber prices for the three species often appear to move in the same direction as each other, with the exception of a few periods in which they are more strongly negatively correlated (e.g., 1950, late 1980s, etc.). However, the degree to which the species' prices actually do move in tandem with each other – measured by the covariance of the percent increase or decrease in prices – depends on which species are being compared. For instance, the species with the greatest covariance in prices are oak and beech – 0.035 (1920-2009). In contrast, the covariance for Monterey pine and beech is less than a quarter of that – 0.008 (1984-2009). Note that neither covariance for these pairs of species is very great, however.

Natural Hazards

Another consideration for landowners in the Basque Country when deciding what species to cultivate is the current blight of pitch canker (*Fusarium circinatum*), a fungus that has infested Monterey pine plantations in particular, killing the trees and commanding lower prices. Whereas Douglas fir and many pine species seem to be relatively resistant to the disease, Monterey pine is much more susceptible (Omsbry, 2004). To address this issue in Spain, the government established a national program to eradicate and control *F. circinatum*. Under the program, landowners and foresters are required to notify the government immediately if they find evidence of pitch cancer in their stands, to destroy any infested stands, and to refrain from planting the area with any species susceptible to *F. circinatum* such as Monterey pine (Ministerio de Agricultura, Pesca y Alimentactión, June 9, 2006).

Forest Policy

Spanish and Basque Country forest policies and subsidies can play a large role in landowners' decisions as to which species of trees to cultivate. For instance, sources in the Basque Country relay that the province of Navarre removed subsidies for Monterey pine a couple of years ago primarily due to the infestation of pitch cancer in these stands; since then, very few stands (if any) of Monterey pine have been planted in Navarre. In contrast, subsidies for Monterey pine plantations continue in the three provinces of Àlava, Gipuzkoa, and Bizkaia. The amounts of the subsidies vary across the three areas but – on average – they amount to about 35 percent of the costs of cultivating pine (e.g., site preparation, cleaning, and pruning costs). Also, sources in the Basque Country report that the Provincial Government of Gipuzkoa recently decided to set aside 1.3€ million to subsidize the harvest of "overgrown" pine plantations that would command little or no value at the sawmill. In general, forest subsidies may come in various forms, change from year to year, and can be difficult to plan for and thus include in financial analyses – especially as they relate to the harvest, which will not occur for many years.

Implications for Forestry Investments

All of the above risks should be taken into account as landowners in the Basque Country – and around the world – decide which tree species to cultivate and whether to plant mixed-species and/or mixed age-class stands. The timber price risk can be incorporated directly into the discount rate for many species via the CAPM, and landowners should at least qualitatively evaluate the other threats to earning a return on the investment and/or managing the forest for a host of other values.

MANAGING TIMBERLAND FOR A DIVERSE SET OF VALUES

"Society is asking for sustainable forestry emphasizing biodiversity and naturalistic forest management" (Spiecker, 2003, p. 57). When considering the financial implications of maintaining monoculture plantations like Monterey pine or transitioning these plantations to mixed-species forests, it is important to keep in mind the effects that this decision may have on water quality, resilience to disease or weather disturbances, biodiversity, carbon sequestration, and the aesthetic quality of the stand. This is especially important for public lands, which are ostensibly being managed for the public good and should take into consideration the various long-term effects of management decisions. While private landowners cannot be expected to absorb financial losses in order to, for example, increase biodiversity, they may take into account how their management regime influences these other values when deciding the species and type of structure they will cultivate. In fact, studies show that non-timber benefits do influence many non-industrial private forest owners' forest management behavior (e.g., Clawson, 1979; Binkley, 1981; Johansson, 1985). While a complete review of the literature on the impacts of different forest and species types on these variables is beyond the scope of the present analysis, various relevant articles are cited for their contribution to our framework for evaluating the ecological implications of transitioning coniferous to broadleaved forests in the Basque Country.

Water Quality

Much attention is paid to how water resources are managed around the world and the ways in which water quality can be maintained at a high level. While Blazewicz (in this volume) provides an investigation of this topic in the Basque Country, it seems relevant to briefly explore this issue here, since landowners should consider the potential effects coniferous species may have on water quality in comparison to deciduous trees like oak and beech.

A noteworthy way in which tree species can influence water quality is through their affects on forest nitrate. Rothe (2005) finds that coniferous forests in central Europe tend to have more than double the concentrations of nitrate than broadleaved forests. Conifers have a higher leaf area that is maintained throughout the year, and this leads to higher rates of nitrogen (N) deposition from the atmosphere. MacDonald et al. (2002) reports that there is a significant relationship between the rates of N deposition and N outputs, concurrent with studies indicating higher inception losses in coniferous than deciduous forests (e.g., Roberts, 1999). Nitrate is highly soluble in water, and the increasing N-saturation of forests has lead to concerns about contamination of groundwater and stream water with nitrate (Rothe 2005).

Water quantity may also be influenced by the extent and type of forest cover present. For instance, Lara et al. (2009) finds a significant positive correlation between native forest cover and annual runoff across 20 watersheds in Chile, as well as a significant negative correlation between summer runoff and percentage of exotic plantations in those watersheds. The highest runoff coefficients (0.42-0.53) occurred in watersheds with about 90% native forest coverage, in contrast to the minimum coefficient (0.30) that was found in the watershed with over 60% young Monterey pine and Eucalyptus plantations (16- and 12-years old, respectively).

Monterey pine plantations may exhibit higher water use than beech or oak forests over the same time frame in part because of the former's shorter rotation length. Studies such as Farley et al. (2005) and Almeida et al. (2007) show that runoff decreases substantially after trees are newly planted and continues to decrease as the trees grow, drawing up more water through their longer roots and allowing more evapotranspiration through their increasingly abundant leaves. This continues for at least twenty years after stands are planted, and then the rate at which the runoff decreases gradually slows. Since pine plantations are replanted every 30-40 years, water use is likely to be higher in these plantations compared with beech or oak forests that take over one hundred years to mature.

Biodiversity

The variety of organisms that naturally occur in a geographic area such as the Basque Country interact with each other in complex ways and – together with their physical environment – form an ecosystem. Because of this interconnectedness, the presence of different species of plants or trees may provide habitat for other species. For instance, it has been postulated that mixed-species forests boast greater biodiversity because they offer diverse habitat for various species (e.g., Peck, 1989; Butterfield and Benitez Malvido, 1992; Elmer et al., 2004). There may be a greater supply of specific food resources, foraging sites, nest locations, or shelter in mixed species forests

(Smith, 1992). Also, wider ecological gradients may be present in mixed-species stands, which in turn may allow for the coexistence of a greater number of species (Simmons and Buckley, 1992).

On the other hand, the effect of mixtures depends greatly on the species or group of species considered, and the impact of stand structure and thus stand management are often greater than the effects of mixed-species stands on biodiversity (Knoke et al., 2008). For example, the structural components of old growth stands such as old logs and hollow trees that may be valuable to certain species often decline or disappear in managed forests (Bengtsson et al., 2000). Studies undertaken in the Basque Country measuring the difference between the numbers of species found in managed Monterey pine plantations as opposed to natural forests with various species – although dominated by beech or oak – reach similar conclusions. For instance, Atauri et al. (2005) finds that the woody understories of deciduous oak and beech forests were significantly richer and more diverse than all of the pine plantations studied (which included plantations aged 5-20 years, over 20 years old, and clearcuts and plantations less than a year old). This corresponds to first-hand observations of managed pine plantations in the Basque Country, which revealed a general paucity of flora and fauna below the canopy. Furthermore, if mixed-species beech and oak forests are fenced to decrease grazing pressure on these trees, other grass and shrub species may also grow in these forests, thereby contributing to further species- and structural-diversity.

Resilience to Natural Hazards

In general, native and thus site-adapted trees are less vulnerable to natural hazards such as windfall, snow breakage and insect attacks (Knoke et al., 2005). For instance, Dieter (2001) finds that survival probabilities for spruce – a non-native species – decrease more sharply than for beech as both trees age. Knoke et al. (2008) discovers that Norway spruce is ten times more susceptible to windthrow than red oak or Pedunculate oak and four times higher than that of European beech This study points out that conifers such as spruce have shallow rooting systems and needle retention during the winter, when storms often occur. Furthermore, Toumey and Korstian (1937, p. 282) describe that "a mixture of shallow-rooted species with deep-rooted species forms a stand that suffers less from wind and more fully utilizes the soil," which offers a possible explanation for the higher resistance to storm damage of mixed coniferous stands with deciduous trees when compared to virtually pure conifer stands shown by Schütz et al. (2006).

Given the increased susceptibility of non-autochthonous species to natural disasters like windthrow and insect infestations – and the likelihood that storms will become more frequent as our planet warms (Lindner et al., 2008) – it follows that resistance to natural disasters like storms and hurricanes would be part of land managers' decision as to which species to cultivate. In fact, Knoke et al. (2005) shows that taking into account the risk of natural hazards in two forests – spruce and beech – narrowed the relation of the financial returns between the two forests from 2.1:1 to 1.7:1. Although it is beyond the scope of the present study, it would be helpful for land managers in the Basque Country if the relative risk of exotic pines versus native beech and oak forests could be assessed and incorporated into a financial model.

Forest and Soil Carbon

As we learn more about the ways human activities influence the levels of carbon dioxide in the atmosphere, forest managers are already taking into account how their own management decisions may positively or negatively affect global carbon dioxide levels. Forest losses, gains, or changes lead to gains or losses in carbon stocks and the levels of carbon dioxide in the atmosphere. Forest ecosystems act as natural stores of carbon, contributing between 50 to 80 percent of terrestrial aboveground carbon storage (see Turner et al., 1995; Kirschbaum, 1996). The general conclusion from the existing literature on carbon storage is that the longer the trees are allowed to grow and sequester carbon, the greater the carbon stock they can store (e.g., Balboa-Murias et al., 2006; Wayburn et al., 2007).

In particular, the "average 100-year cycling time of carbon [is] the desired time frame in which to operate" (Wayburn et al., 2007, p. 9). Of course, 30- to 35-year rotations for Monterey pine plantations are much shorter rotation than this, although beech and oak have rotation lengths in excess of one hundred years. Cooper (1983) and Cannell (1999) use logistic growth curves to model forest carbon accumulation, and they conclude that the time-averaged carbon storage from planting to harvest (at the time of maximum Mean Annual Increment) is about a third of the carbon stored in the forest at full maturity. Additionally, Jandl et al. (2007) reports that trees with high wood density (many broad-leaved tree species) accumulate more carbon than trees with light wood (many coniferous tree species). Thus, it can be concluded that longer rotation-length species such as beech and oak have a greater positive impact on carbon storage than shorter-rotation species such as Monterey pine.

In addition to aboveground biomass, soil organic matter is an important carbon sink. Global estimates of the amount of carbon stored belowground vary, but it is generally accepted that soil contains 2-3 times more carbon than biomass (Post et al., 1990; Eswaran et al., 1993; Batjes, 1996; Schlesinger, 1997; Sheikh, Kumar, and Bussmann, 2009). Different ecosystems and soils store more or less carbon than others – e.g., boreal forests store more carbon in detritus than do tropical forests. In temperate forests like those found in the Basque Country, the ratio of aboveground to belowground carbon storage is about 6:5 (Schlesinger, 1997).

The pool of carbon contained belowground is determined by a balance of carbon input from 1) litterfall and that released by roots into the soil, and 2) carbon released from decomposition. Various studies have found that forest management activities have different effects on carbon sequestration in soils (Johnson, 1992; Post and Kwon, 2000; Johnson and Curtis, 2001), and Jandle et al. (2007) gives a comprehensive review of the literature on this topic. Gartzia-Bengoetxea et al. (2009) reports that mechanical site preparation and harvesting in Monterey pine plantations in the Basque Country result in a significant loss of organic matter that may persist for at least 16 years after the disturbance. In general, intensive harvest techniques that substantially disturb the soil and that leave few leftover branches/leaves from the harvest may cause more carbon to be released into the atmosphere than any additional carbon input from leftover residues.

As the global community looks towards ways of reducing carbon dioxide emissions and carbon markets expand, the price of carbon may increase sufficiently so as to make the different carbon sequestration volumes in the two stand types another important factor in such a decision (if not *the* most important factor). Further exploration of the potential earning power of such trees via carbon markets would contribute to any analysis on species and stand management decisions.

Conclusion

The results of this paper do not support the financial superiority of cultivating mixed-species beech and oak forests in comparison to single-species Monterey pine or other conifer species in the Basque Country. The much longer rotation lengths of the latter species make it virtually impossible to achieve financially more attractive forests populated by beech and oak, barring a drastic, long-term drop in pine prices or such an increase in beech and/or oak prices.

Other important considerations should – and most likely are – taken into account by land managers when deciding between such alternatives as the ones presented above. Factors such as financial risk, ecological resilience, and effects on water quality, biodiversity, and carbon sequestration should all be considered in forest management decisions. Determining which factors are most important in such decisions depends on the strategic goals of the landowner, but financial risk and structural resilience could have the largest impact on the financial value of the decision choices presented and thus may provide the most compelling argument from a financial standpoint.

Going forward, several contributions could be made to further this analysis. First, research on the market in the Basque Country for oak and beech timber would be helpful. It was assumed in this analysis that the prices for the species in Spain would be quite similar to those in France, but there are several caveats to be explored related to this assumption. For instance, given the difference in the types of mills and equipment that would be able to process softwoods vs. hardwoods, oak and beech may need to be transported to other regions of Spain or France (such as Navarre) for processing, which introduces questions of taxation and higher costs of transportation. Of course greater supply of these species may also cause prices to fall somewhat, depending on the supply from the rest of the world and whether such species would be exported or bound primarily for domestic markets.

Second, as mentioned towards the beginning of the paper, there are various alternative species that landowners may find financially rewarding to cultivate that have not been explored in this paper – ash, walnut, and cherry, to name a few. If one of these hardwood species were planted along with a fast-growing but resistant softwood species, both financial reward and risk reduction may be achieved.

Finally, an analysis presenting the financial benefits of choice in investment decisions would also be helpful. For instance, cultivating a species with a shorter rotation length allows the landowner more flexibility in deciding which species she/he will cultivate after harvest of the present species. The ability to manage adaptively and respond to changes in the market presents advantages that may be incorporated into the financial analysis.

REFERENCES

Agirre, J. A. (2009). Análisis económico sobre la aplicación de la poda en pino insignis (pinus radiata) para la producción de madera libre de nudos. *Euskadi Forestal,* 78, 36-43.

Almeida, A. C., Soares, J. V., Landsberg, J. J. and Rezende, G. D. (2007). Growth and water balance of Eucalyptus grandis hybrids plantations in Brazil during a rotation for pulp production. *Forest Ecology and Management*, 251, 10-21.

Atauri, J. A., De Pablo, C. L., De Agar, P. M., Schmitz, M. F. and Pineda, F. D. (2005). Effects of forest management on understory diversity in the forest ecosystems of northern Spain. *Environmental Management,* 35, 819-828.

Balboa-Murias, M. A., Rodríguez-Soalleiro, Merino, R. and Álvarez-González, J. G. (2006). Temporal variations and distribution of carbon stocks in aboveground biomass of radiata pine and maritime pine pure stands under different silvicultural alternatives. *Forest Ecology and Management,* 237, 29-38.

Batjes, N. H. (1996). Total carbon and nitrogen in soils of the world. *European Journal of Soil Science,* 47, 151-163.

Beltratti, A., Chichilnisky, G. and Heal, G. (1996). Sustainable Use of Renewable Resources. Social Science Research Network Electronic Library. Retrieved January 29, 2010 from http://papers.ssrn.com/abstract=1613.

Bengtsson, J., Nilsson, S. G., Franc, A. and Menozzi, P. (2000). Biodiversity, disturbances, ecosystem function and management of European forests. *Forest Ecology and Management,* 132, 39-50.

Binkley, C. S. (1981). Duration dependent discounting? *Canadian Journal of Forest Research,* 11, 454-456.

Brazee, R. J. and Mendelsohn, R. (1988). Timber Harvesting with Fluctuating Prices. *Forest Science,* 34 (2), 359-372.

Butterfield, J. and Benitez Malvido, J. (1992). Effect of mixed-species tree planting on the distribution of soil invertebrates. In Cannell, M. G. R., Malcolm, D. C. and Robertson, P.A. (Eds.). *The ecology of mixed-species stands of trees.* Oxford: Blackwell.

Cannell, M. (1999). Environmental impacts of forest monocultures: water use, acidification, wildlife conservation, and carbon storage. *New Forests,* 17, 239-262.

Caulfield, J. P. (1998). Timberland in institutional portfolios and the question of persistence. *Forest Products Journal,* 48, 23-28.

Chauchard, L. M. (2000). Crecimiento y producción de repoblaciones de *Pinus radiata* D. Don en la provincia de Gipúzkoa (País Vasco). Doctoral dissertation, Escuela Tecnica Superior de Ingenieros de Montes, Madrid, Spain.

Clawson, M. (1979). *Economics of US non-industrial private forests.* Washington, DC: Resources for the Future.

Cooper, C. F. (1983). Carbon storage in managed forests. *Canadian Journal of Forest Restoration,* 13, 155-166.

De Vries, W., Reinds, G. J., Posch, M., Sanz, M., Krause, G. and Calatyud, V. et al. (2003). Intensive Monitoring of Forest Ecosystems in Europe. *TeChapter Rep*., EC. UN/ECE, Brussels.

Dieter, M. (2001). Land expectation values for spruce and beech calculated with Monte Carlo modeling techniques. *Forest Policy and Economics,* 2, 157-166.

Elmer, M., La France, M., Förster, G. and Roth, M. (2004). Changes in the decomposer community when converting monocultures to mixed Norway spruce/European beech stands. *Plant and Soil,* 264, 97-109.

Eswaran, H., Van Den Berg, E. and Reich, P. (1993). Organic carbon in soils of the world. *Soil Science Society of American Journal*, 57, 192-194.

Euskadi Forestal. Septiembre 2009. Retrieved January 25, 2011 from http://basoa.org/revista-completa/.

Fagus Sylvatica Fichero. Retrieved January 29, 2010 from http://www.basoa.org/selvicola.asp?id=308andpadre=10.

Farley, K. A., Jobbagy, E. G. and Jackson, R. B. (2005). Effects of afforestation on water yield: a global synthesis with implications for policy. *Global Change Biology,* 11, 1565-1576.

Forrester, D. I., Bauhus, J. and Cowie, A. (2005). On the success and failure of mixed-species tree plantations: lessons learned from a model system of *Eucalyptus globulus* and *Acacia mearnsii. Forest Ecology and Management,* 209 (1/2), 147-155.

Gartzia-Bengoetxea, N., González-Arias, A. and Martínez de Arano, I. (2009). Effects of tree species and clear-cut forestry on forest floor characteristics in adjacent temperate forests in northern Spain. *Canadian Journal of Forest Research,* 39, 1302-1312.

Haight, R. G. (1985). A comparison of dynamic and static economic models of uneven-aged stand management. *Forest Science,* 31 957-974.

Haight, R. G. (1990). Feedback Thinning Policies for Uneven-Aged Stand Management with Stochastic Prices. *Forest Science,* 36 (4), 1015-1030.

Hanewinkel, M. (2001). Economic aspects of the transformation from even-aged pure stands of Norway spruce to uneven-aged mixed stands of Norway spruce and beech *Forest Ecology and Management,* 151, 181-193.

Hanewinkel, M. (2002). Comparative economic investigations of even-aged and uneven-aged silvicultural systems: a critical analysis of different methods. *Forestry,* 75 (4), 473-481.

Inventario Forestal CAE (2005). Departamento de Medio Ambiente, Planificacion Territorial, Agricultura y Pesca. Gobierno Vasco.

IKT SA (2002). Actualización de las tablas de producción de Pinus radiata en el País Vasco (Updated production tables for pinus radiata in the Basque Country). This is a continuation of the results of Chauchard (2001).

Jandle, R., Lindner, M., Vesterdal, L., Bauwens, B., Baritz, R. and Hagedorn, F. et al. (2007). How strongly can forest management influence soil carbon sequestration? *Geoderma,* 137, 253-268.

Johansson, P. O. (1985). *The Economics of Forestry and Natural Resources.* Oxford, England: Blackwell Scientific.

Johnson, D. W. (1992). Effects of forest management on soil carbon storage. *Water, Air, and Soil Pollution*, 83-121.

Johnson, D. W. and Curtis, P. S. (2001). Effects of forest management on soil C and N storage: meta analysis. *Forest Ecology and Management,* 140, 227-238.

Kelty, M. J. (2006). The role of species mixtures in plantation forestry. *Forest Ecology and Management,* 233, 195-204.

Kirschbaum, M. U. F. (1996). The carbon sequestration potential of tree plantations in Australia. In: Eldridge, K. G., Crowe, M. P. and Old, K. M. (Eds.). *Environmental Management: The Role of Eucalypts and Other Fast Growing Species*. CSIRO Forestry and Forest Products, 77-89.

Klemperer, W. D. (1996). Forest Resource Economics and Finance. New York: McGraw-Hill.

Knoke, T. (1999). Economic analysis of the wood production in a mixed, uneven-aged forest. In: Olsthoorn, A. F. M., Bartelink, H. H., Gardiner, J. J., Pretzsch, H. and Franc, A. (Eds.) (1999). Management of mixed-species forest: silviculture and economics. Wageningen (Netherlands): Instituut voor Bos- en Natuuronderzoek DLO.

Knoke, T., Moog, M. and Plusczyk, N. (2001). On the effect of volatile stumpage prices on the economic attractiveness of a silvicultural transformation strategy. *Forest Policy and Economics* 2, 229-240.

Knoke, T., Stimm, B., Ammer, C., and Moog, M. (2005). Mixed forests reconsidered: A forest economics contribution on an ecological concept. *Forest Ecology and Management,* 213, 102-116.

Knoke, T., Ammer, C., Stimm, B. and Mosanl, R. (2008). Admixing broadleaved to coniferous tree species: a review on yield, ecological stability and economics. *European Journal of Forest Research,* 127, 89-101.

Lara, A., Little, C., Urrutia, R., McPhee, J., Álvarez-Garretón, C. and Oyarzún, C. et al. (2009). Assessment of ecosystem services as an opportunity for the conservation and management of native forests in Chile. *Forest Ecology and Management,* 258, 415-424.

Lindner, M., Garcia-Gonzalo, J., Kolström, M., Green, T., Reguera, R. and Maroschek, M. (2008). Impacts of Climate Change on European Forests and Options for Adaptation. Report to the European Commission Directorate-General for Agriculture and Rural Development. AGRI-2007-G4-06.

Retrieved July 1, 2010 from: http://ec.europa.eu/agriculture/analysis/external/euro_forests/full_report_en.pdf.

Lohmander, P. (2000). Optimal sequential forestry decisions under risk. *Annals of Operations Research,* 95, 217-228.

MacDonald, J. A., Dise, N. B., Matzner, E., Armbruster, P., Gundersen, P. and Forsius, M. (2002). Nitrogen input together with ecosystem nitrogen enrichment predict nitrate leaching from European forests. *Global Change Biology,* 8, 1028-1033.

Madrigal, A., Puertas, F., Martinez, J., Zazo, J., Arrechea, E. And Eraso, E. et al. (1989). Tablas de producción de selvicultura variable de Fagus sylvatica L. en Navarra. Editorial: Departamento de Agricultura, Ganadería y Montes. Pamplona, Spain.

Markowitz, H. (1952). Portfolio selection. *Journal of Finance,* 7, 77-91.

Mendell, B. C. (Ed.) (2006). Discount rates and timberland investment. Timberland Report, 8 (3). James W. Sewall Company.

Michie, B. R. (1985). Uneven-aged stand management and the value of forest land. *Forest Science,* 31, 116-121.

Ministerio de Agricultura, Pesca y Alimentactión (June 9, 2006). REAL DECRETO 637/2006, de 26 de mayo, por el que se establece el programa nacional de erradicación y control del hongo Fusarium Circinatum

Niremberg et O'donnell. *Boletín Oficial del Estado (BOE), 137*, 22069-22073. Retrieved July 1, 2010 from http://www.boe.es/boe/dias/2006/06/09/pdfs/A22069-22073.pdf.

Mohr, B. and Shori, C. (1999). Femelschlag oder Plenterung – ein Vergleich aus betriebswittschaftlicher Sicht. *Swiss Forestry Journal,* 150, 49-56.

MSCI World Index. Retrieved July 11, 2010 from http://www.mscibarra.com/products/indices/international_equity_indices/gimi/stdindex/performance.html.

Nichols, J. D., Bristow, M. and Vanclay, J. K. (2006). Mixed-species plantations: prospects and challenges. *Forest Ecology Management,* 233, 383-390.

Niese, J. N. and Strong, T. F. (1992). Economic and tree diversity tradeoffs in managed northern hardwoods. *Canadian Journal of Forestry Research,* 22, 1807-1813.

Norstrom, C. J. (1975). A stochastic model for the growth period decision in forestry. *The Swedish Journal of Economics,* 77 (3), 329-337.

Olarieta, J. R., Besga, G., Rodriguez, R., Uson, A., Pinto, M. and Virgel, S. (1999). Sediment enrichment ratios after mechanical site preparation for Pinus radiata plantation in the Basque Country. *Geoderma,* 93 (3-4), 255-267.

Omsbry, M. (2004). Report on the interception of *Fusarium circinatum* (Pitch Canker) on imported seedlings of Douglas fir (*Pseudotsuga menziesii*). MAF Biosecurity New Zealand. Retrieved June 27, 2010 from http://www.biosecurity.govt.nz/files/pests/pitch-canker/pine-pitch-canker-interception.pdf.

Otazua, F. (1998). Invertir en el monte. Revista de AITIM (Asociación de Investigación Técnica de Las Industrias de la Madera). Retrieved November 26, 2009 from http://infomadera.net/uploads/articulos/archivo_3308_11686.pdf?PHPSESSID=4359c7d498f9da78a4639c2778c1d72b.

Peck, K. M. (1989). Tree species preferences shown by foraging birds in forest plantations in northern England. *Biological Conservation,* 48, 41-57.

Pino Insignis Fichero. Retrieved December 10, 2009, from http://www.basoa.org/selvicola.asp?id=308andpadre=10.

Post, W. M., Peng, T. H., Emmanuel, W. R., King, A. W., Dale, V. H. and De Angelis, D. L. (1990). The global carbon cycle. *American Science,* 78, 310-326.

Post, W. and Kwon, K. (2000). Soil carbon sequestration and land-use change: processes and potential. *Global Change Biology,* 6, 317-328.

Prestemon, J. P. and Holmes, T. P. (2000). Timber price dynamics following a natural catastrophe. *American Journal of Agricultural Economics,* 82, 145-160.

Quercus Robur Fichero. Retrieved January 29, 2010 from http://www.basoa.org/selvicola.asp?id=308andpadre=10.

Revue forestière française (1920-2009). Les ventes de bois des forêts publiques. Office national des forêts. Retrieved January 29, 2010 from http://documents.irevues.inist.fr/handle/2042/4752.

Roberts, J. (1999). Plants and water in forests and woodlands. In A. J. Baird and R. L. Wilby (Eds.). *Ecohydrology*. London: Routledge.

Ross, S. A., Westerfield, R. W. and Jaff, J. (2008). *Corporate Finance, 8th Edition*. New York: McGraw-Hill/Irwin.

Rothe, A. (2005). Tree species management and nitrate contamination of groundwater: a central Europe perspective. In: Binkley, D. and Menyailo, O. (Eds.). *Tree species effects on soils: Implications for Global change*. Netherlands: Springer.

Schlesinger, W. H. (1977). Carbon balance in terrestrial detritus. *Annual Review of Ecology and Systematics,* 8, 51-81.

Schmitz, M. F., Atauri, J. A., De Pablo, C. L., De Agar, P. M., Rescia, A. J. and Pineda, F. D. (1998). Changes in land use in northern Spain: effects of forestry management on soil conservation. *Forest Ecology and Management,* 109, 137-150.

Schütz, J. P., Götz, M., Schmid, W. and Mandallaz, D. (2006). Vulnerability of spruce (Picea abies) and beech (Fagus sylvatica) forest stands and consequences for silviculture. *European Journal of Forest Research,* 125, 291-302.

Sheikh, M. A., Kumar, M. and Bussmann, R. W. (2009). Altitudinal variation in soil organic carbon stock in coniferous subtropical and broadleaf temperate forests in Garhwal Himalaya. *Carbon Balance and Management,* 4:6. Available at: http://www.cbmjournal.com/content/4/1/6.

Scott, C. W. (1960). Radiata Pine as an Exotic. *Unasylva,* 14 (1). Retrieved November 14, 2009 from the FAO Corporate Document Repository http://www.fao.org/DOCREP/X5394E/X5394E03.HTM.

SECOMA (Servicios Comerciales de la Madera de Gipuzkoa). The Industry: Pine Process. Retrieved November 22, 2009 from http://www.secoma.org/Procesociclo.php.

Sharpe, W. F. (1964). Capital asset prices: a theory of market equilibrium under conditions of risk. *Journal of Finance,* 14, 425-442.

Simmons, E. A. and Buckley, G. P. (1992). Ground vegetation under planted mixtures of trees. In: Cannell, M. G. R., Malcolm, D. C. and Robertson P. A. (Eds). *The ecology of mixed-species stands of trees.* Oxford: Blackwell.

Smith, K. W. (1992). Bird populations: effects of tree species mixtures. In Cannell, M. G. R., Malcolm, D. C. and Robertson P. A. (Eds). *The ecology of mixed-species stands of trees.* Oxford: Blackwell.

Spiecker, H. (2003). Silvicultural management in maintaining biodiversity and resistance of forests in Europe-temperate zone. *Journal of Environmental Management,* 67, 55-65.

Tarp, P., Helles, F., Holdern-Anderson, P., Larsen, J. B. and Strange, N. (2000). Modeling near-natural silvicultural regimes for beech – an economic sensitivity analysis. *Forest Ecology and Management,* 130, 187-198.

Toumey, J. W. and Korstian, C. F. (1937). Foundations of Silviculture upon an Ecological Basis. New York: John Wiley and Sons.

Thomson, T. A. (1991). Efficient Combinations of Timber and Financial Market Investments in Single-Period and Multiperiod Portfolios. *Forest Science,* 37 (2), 461-480.

Turner, D. P., Koerper, G. J., Harmon, M. E. and Lee, J. J. (1995). A carbon budget for forests of the conterminous United States. *Ecological Applications,* 5 (2), 421.

Wagner, J. E. and Rideout, D. B. (1991). Evaluating forest management investments: the Capital Asset Pricing Model and the Income Growth Model. *Forest Science,* 37 (6), 1591-1604.

Washburn, C. L. and Binkley, C.S. (1990). Informational Efficiency of Markets for Stumpage. *American Journal of Agricultural Economics,* 72 (2), 394-405.

Wayburn, L. A., Franklin, J. F., Gordon, J. C., Binkley, C. S., Mladenoff, D. J. and Christensen, N. L. (2007). Forest carbon in the United States: opportunities and options for private lands. A report for the Pacific Forest Trust. Retrieved January 29, 2010 from http://www.pacificforest.org/policy/forestcarbon.html.

Wood Product Prices: Report from Europe, the UK, and Russia (October 16-31st, 2009). Retrieved December 9, 2009 from www.globalwood.org/market/timber_prices_2009/aaw1002e.htm.

Valkonen, S. and Valsta, L. (2001). Productivity and Economics of mixed, two-storied spruce and birch stands in southern Finland simulated with empirical models. *Forest Ecology and Management,* 140, 133-149.

Valsta, L. T. (1998). Economy in the short- and long-term perspectives. New stand types in Boreal forestry. Finnnish Forest Research Institute research paper 714. (As cited in Knoke et al. 2001.).

Zerbe, S. (2002). Restoration of natural broad-leaved woodland in Central Europe on sites with coniferous forest plantations. *Forest Ecology and Management,* 167, 27-42.

Zinkhan, F. C., Sizemore, W. R., Mason, G. G. and Ebner, T. J. (1992). *Timberland investments: a portfolio perspective*. Portland: Timber Press.

In: Multiple-Use Management ... ISBN 978-1-61942-328-2
Eds: A. DuVal, T. Muruetagoiena et al.

Chapter 5

Management Approaches for Protected Areas of Cultural and Biological Significance: The Beech Pollards of Aiako Harria Natural Park

Stella J. M. Cousins and Ashley E. DuVal*
Yale School of Forestry and Environmental Studies,
New Haven, CT, US

Abstract

The Basque Country is home to Europe's largest remaining stands of pollarded beech forest. Within Aiako Harria Natural Park in the Gipuzkoa Province of Spanish Basque Country, an estimated 800 ha of pollarded beech (*Fagus sylvatica* L.) forest compositionally dominate the landscape. These trees were pollarded continuously over centuries to provide ship parts and charcoal for Basque industry. Recent studies have supported the role of senescing pollards, abundant sources of standing dead wood, in promoting local biodiversity through habitat creation and a food source for saproxylic invertebrate species. Despite this recent revival in interest, there remains limited public awareness of the cultural and ecological importance of pollards. Park managers are at a critical point in

* Corresponding authors: stella.cousins@alumni.yale.edu, ashley.duval@yale.edu.

administering maintenance to promote the long-term survival of these culturally modified trees, which are threatened by structural instability resulting from a lapse in management. Building upon a historical and ecological overview, this paper draws from our observations in Aiako Harria as well as other case studies to highlight important considerations for protected area managers in promoting and managing landscapes of cultural and ecological importance. Suitable best practice approaches include an inventory and scientific study, increased outreach and education to build greater public awareness, a multiple use management approach, and integration of cultural and biological frameworks for valuing the pollard forests.

Keywords: pollard, *Fagus sylvatica*, Basque country, biodiversity, saproxylic, veteran trees, cultural heritage, managed landscape, culturally modified tree

Introduction

Aiako Harria Natural Park, typical of the Spanish Basque country, is characterized by landscapes altered by the long legacy of the Basque people and their utilitarian relationship with the environment. Throughout the region, the signature of settlement and management is evident in a mosaic of centuries-old pasturelands and contemporary conifer plantations, but the landscape also bears more subtle hints from ancient times. Roman mines and the early port at Irun evidence land use and resource extraction dating back at least to the first century BC (Barandiarán, 1973; Benito, 1988). Within the park, stone circle monuments known as *cromlechs* recall settlements from 1000-600 BC (Altuna and Areso, 1977), and single-chamber megalithic tombs called *dolmens* from the neolithic age suggest a history of pastoralism in the mountains of Aiako Harria since 3000 BC. The steep pastures were cleared by shepherds more than 4,000 years ago, when the pedunculate oak (*Quercus petraea* (Matt.) Liebl.), rare on the landscape today, was a dominant species. Forests dominated by beech (*Fagus sylvatica* L.) became common in the area 3,500 years ago (Elosegi, personal communication 3/17/10). While stone and structural artifacts provide obvious indications of prehistoric occupation, the extensive record of Basque settlement is evidenced by clues in the natural landscape as well. Centuries of grazing, cultivation, and land clearing have transformed the region's hardwood forests into today's characteristic high elevation grasslands and heath. In Aiako Harria Natural Park and many other locations in the region, extensive tracts of forest are compositionally dominated by pollarded beech and oak trees managed for hundreds of years as a renewable source of fuel, timber and fodder. These culturally modified trees represent a threatened form of traditional ecological knowledge as only a handful of aging individuals hold firsthand experience in their management. As both important cultural artifacts and living components of a dynamic ecological system, the pollards pose unique management challenges. There have been significant changes to forest structure since the lapse of routine pollarding. The dense, closed beech canopy has shaded out advanced regeneration resulting in a single uniform age class that dominates the forest. With many of the pollards nearing or surpassing the traditional time between cutting cycles, the stands in the park today are top-heavy and prone to breaking in windstorms or under their own weight. The loss of many larger pollarded trees in recent years underscores the necessity of management interventions to secure them for future generations.

The daunting size and gnarled forms of the pollarded trees set against a closely grazed understory provide arresting landscapes that attract hikers and picnickers to the park. However, for those who frequent the mountains of the Basque Country, the presence of these gnarled trees is ubiquitous and often unnoticed. In a region renowned for 2,000 year old Roman mines, 3,000 year old cromlechs and cave art dating back 18,000 years, the historical value of pollarded trees that represent over 200 years of traditional management have largely escaped notice. Recently, a visit of ancient tree enthusiasts on a European tour of pollarding practices drew international recognition to the trees as a unique and threatened cultural heritage (Read, 2006). Teams of resource managers, arborists and biologists have since conducted studies on the regeneration of maiden (not pollarded) beech forests, the biodiversity and ecological state of these stands, and local knowledge of traditional pollarding techniques used by charcoal makers (General Directorate for Woodlands and the Natural Environment, 2005). Recently, the Woodland Trust, a British organization focusing on ancient and veteran tree preservation, has expressed interest in how data gained from the study of these pollards could inform their own pollard and veteran tree management practices and vice versa (Read, 2007).

This analysis seeks to address the underlying management issues posed by a biocultural landscape through a review of comparable systems where some success managing protected areas for mixed values has been achieved. We begin with a historical overview leading to a discussion of the present status of the pollarded stands, current management practices, and their associated values.

THE POLLARDED BEECH FORESTS

History

Aiako Harria Natural Park contains more than 800 hectares of pollarded forests, composed primarily of European beeches and, less commonly, pedunculate oak (*Quercus robur* L.), European ash (*Fraxinus excelsior* L.), and Iberian white birch (*Betula celtiberica* Rothm. and Vasc.). These pollarded beech forests in the mountainous Basque region comprise perhaps the largest extant aggregations of pollard-managed forest left in Europe. The tradition of pollarding has an long history of practice in Europe, where fossil oak pollards discovered near the River Trent in the UK were dated at more than 3,400 years old (Butler, 2006). As a silvicultural and silvo-pastoral technique, pollarding was once widespread in Europe but is relatively uncommon today (Petit and Watkins, 2003).

In Gipuzkoa Province, pollarding became common on a large scale during the 16th and 17th centuries when improved technologies for melting iron and crafting ships coupled with population growth drove up the demand for forest products. Due to the region's abundant mineral resources and water, the demand for iron was met by foundries and furnaces stocked with charcoal from nearby forests. The region transitioned to large-scale use of iron ore with the advent of industrial iron melting through river foundries, marking a change in the processing capacity from the small melting ovens operating during the Middle Ages (Elosegi, personal communication 3/17/10). At the same time, the needs of merchants, fishermen, and the Spanish Navy were rapidly expanding to serve newfound trade routes. Shipbuilders could pay for curved timbers for ship ribs and needed strong, Y-shaped timbers for critical joints, driving demand for pollarded forms (Rozas, 2004; Aragon Ruano, 2001). As the expansion of local enterprises drove population growth in the region, the beech pollard areas were managed for the fodder and nut crops needed to sustain livestock. Free range grazing of sheep, horses, and cattle was common throughout the region, following traditional routes of transhumance and extensively utilizing forested areas. Pigs were also released on a seasonal basis to consume mast (Rozas, 2003; Loidi, 2005).

Thus, pollard management developed in the region along two product-focused pathways, while accommodating pastoral uses. Trees to be sent to sea were carefully selected and topped at about 3m above ground to encourage horizontal growth of two to three selected branches. Called *guiados,* the branches of these trees were typically guided with weights and careful attention to perfect their form. Cutting and shaping of other branches would occur every 10-12 years (slightly more often for beech trees) until the development of the primary branches was complete 70-100 years later (Aragon Ruano, 2001). Each visit would include removal of the smaller sprouts since the last cutting. In the end, the entire tree would be harvested for timber. In contrast, charcoal production provided a more frequent harvest and flexibility of use. These pollards, known as *trasmochos,* were typically cut back using axes every 10-20 years, removing some or all of the branches from the single trunk, similarly at about 3m above ground. Straighter stems of small diameter were ideal for use in the furnaces for manufacturing iron and in limekilns, as well as for domestic purposes. These two types of management would have likely intermixed, with more accessible locations favored for *guiados* and the steeper, more remote terrain utilized solely for *trasmochos* (Fenley, 1950; Aragon Ruano, 2001; Loidi, 2005; Read, 2006). New methods of pollarding were developed out of the necessity for multiple uses, and quickly proved to be an effective means of satisfying resource demands. In 1548, the government of Gipuzkoa recognized their importance for the people of the region, and mandated that more trees be pollarded (Aragon Ruano, 2001; Read, 2004).

Over the next two hundred years, demand for wood for in the Basque region continued to increase, and forested areas dwindled. Lowland forests were displaced by agriculture as populations expanded, and the demand for wood grew (Kenk and Guehne, 2001; Onaindia and Mitxelena, 2009). Forests in municipal control, such as those that now comprise 83% of Aiako Harria Natural Park, were somewhat protected from these pressures by virtue of their status as public utility property. Some of the most frequented pollarded sites in the province, such as Oieleku and Artikutza, retained native forest and pollarding activities as they were poorly suited for farming and distant from town centers. The 20th century saw further changes in land use, with widespread agriculture abandonment and reforestation with exotic species. As demand for traditional wood products such as charcoal and structural timbers declined drastically, forestry practices transitioned towards management for paper pulp, and still more native forests were converted to plantations of exotic conifer species (Garcia-Ruiz et al., 1996; Rescia et al., 1997, Schmitz,

1998; Loidi, 2005). In the second half of the 20 th century, regular cutting of beech and oak in the high hills became too labor intensive to warrant establishment of new pollards on a commercial scale. This followed a period of decline in charcoal making, which was abandoned altogether in the 1960s with the exception of the regions of Urbasa and Lokiz in Navarre where the tradition continues, however limited (A. Elosegi, personal communication 16 March 2010). Many pollarded forests were removed and replaced with plantations, and while thousands of hectares remain in the mountains of the Basque Country under different forms of land ownership, only a few hundred hectares remain within the protection and jurisdiction of Aiako Harria Natural Park.

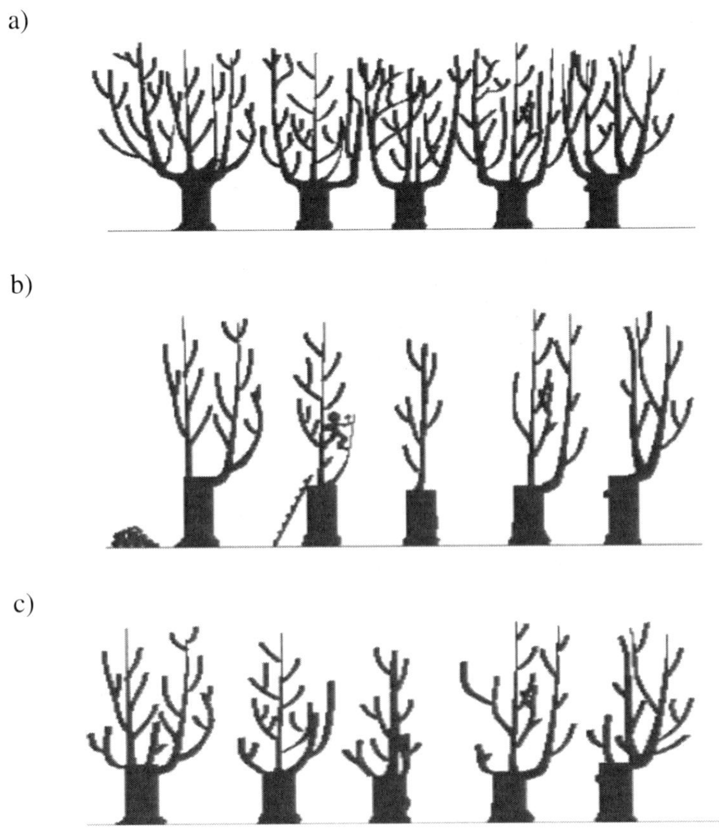

Figure 1. The pollarding management cycle. A. Pollarded trees one year prior to cutting; B. immediately following a cutting, and; C. one to a few years following a cutting, with branches resprouting from the bole. Adapted from Loidi, (2005).

Present Condition

In the park today, the pollards provide habitat for a variety of species but they present interesting and urgent management challenges due to their continuously changing condition. Some of the senescing individuals have already succumbed to disease and decay, lost limbs in storm events or simply split and collapsed under their own immense weight. This process provides a pulse of coarse woody debris to the forest floor and its inhabitants (Figure 1B, 1C). In areas free of grazing, a significant threat to the relict pollarded trees also comes from competition and shading by uncut, or maiden, beeches. The process of tree death and decay may be ecologically ephemeral, but it nevertheless provides a rich habitat for species characteristic of mature forests. The trees, often hundreds of years old, have attributes such as nest holes, soft wood, and woody debris and serve as a refugia to a suite of species uncommon in Europe by virtue of the rarity of these habitats (Franklin et al, 1987; Ódor and Standovár, 2001; Grove, 2002; Christensen et al, 2004). Although the pollarded beech forests were once thought of as "biodiversity deserts," recent surveys of the plant and animal communities found in the park have revealed diverse assemblages of insects, lichens, and fungi established upon the decaying wood of the senescing trees (General Directorate for Woodlands and the Natural Environment, 2005). A new species of lichen was documented in 2008, along with observations of 57 species of lichens and lichenicolous fungi documented in the Park for the first time in the Basque country, and some for the first time in continental Europe. Pollard forests support a wide diversity of invertebrates; at least 268 species of insects in 68 families are found in the Oieleku area, 109 of which are saproxylic (species consuming and inhabiting wood). The cerambycid *Rosalia alpina*, a European Protected Species, occurs in the decaying woods (Carte, 2008; Etayo, 2008). The recent LIFE project stresses protection of habitat for this species as well as *Lucanus cervus* and *Osmoderma eremita,* two other protected beetles that rely on rotting wood. Some of these endangered species combine xylophagous larvae with adults who feed on grass or pollen, necessitating the combined management of dead wood with open glades. Within the past decade, Park managers have embarked on a series of investigations focused on the beech forests with perpetuation of this biodiversity in mind.

A number of livestock exclusion plots were established to secure regeneration of woody species (Figure 1D) and former plantation sites have been replanted with beech saplings (General Directorate for Woodlands and the Natural Environment, 2005).

Figure 2. Status of decay and regeneration in the pollard forests of the Basque Country. A. Pollarded beech forest with grazed understory, Oieleku recreation area. B. Senescing individuals provide abundant woody debris for the forest floor, Artikutza. C. Increasingly heavy branches on abandoned individuals cause eventual collapse, Artikutza. D. In an experimental exclosure, beech seedlings and other understory vegetation fill the gap created by a recently collapsed pollard. Near Oieleku. E. Use of traditional methods to reinstate pollarding for a beech at Urkizu. Images courtesy A, B, C, D.: S. Cousins; E.: H. Read.

Pollarded trees may live for twice the typical lifespan of an unmodified forest beech because of management, but these will advance into structural instability in the next 10-20 years if not carefully pruned or returned to a pollarding system (Read, 2006; A. Elosegi, personal communication, October 21 2009). In a few locations, the trees are just barely out of their traditional cutting cycle. With the susceptibility of top-heavy pollards to storms and decay, intervention is essential to ensure the future of these forests. One of the outcomes of the collaboration between the Woodland Trust and the Basque Biodiversity Center has been a trial of both modern and traditional cutting techniques for pollarded trees (as seen in Figure 2E). The results of this study are being monitored and are expected to provide important information on the suitable methods and costs associated with continuing pollard management in some areas of the park (Read, 2007). Through the LIFE project focused on Aiako Harria Natural Park, 27 conservation actions have been completed, the majority of which focused on native hardwood forests. The program concentrates efforts on the restoration and conservation of habitats and species of community interest through innovative methods, and stresses demonstrational and educational aspects. Park managers have also recently secured a LIFE+ grant for the period 2010-2013, which is focused primarily on the beech pollards (Segura, 2009).

KEY QUESTIONS

The current condition of Aiako Harria's pollard forests poses a number of management concerns. The most pressing is to determine how transitional landscapes should be managed as senescing trees are replaced by forests that differ in structure and composition. A second concern is the provision of long-term management, which is costly and labor-intensive, and no longer driven by the need for ship parts or fuel.

The third and most fundamental challenge is that of balancing biological and cultural values in protected area management. Every forest type and management approach carries a different set of conservation values and trade-offs. The pollards can be valued for their role in conserving both local biodiversity or cultural heritage. Management in line with these different values encompasses distinct mechanisms for valuation, actions, and financial support. Conservation groups, NGOs, and various international organizations increasingly recognize the value of applying an inclusive and holistic approach to managing areas of natural and cultural importance. These approaches should be considered as a possible theoretical basis for future management decisions taken in the park.

CASE STUDIES

What follows is an abbreviated review profiling a range of approaches to management and preservation of sites that have both ecological and cultural importance. The case studies that follow showcase approaches to the management of biocultural landscapes and resources under a wide range of circumstances with the goal of illuminating potential management paths for both the beech pollards at Aiako Harria and for other sites with multifaceted biocultural values.

BURNHAM BEECHES AND COUNTRYSIDE MANAGEMENT IN THE UNITED KINGDOM

Burnham Beeches

The Burnham Beeches is a National Nature Reserve and Site of Special Scientific Interest owned by the City of London and located in Buckinghamshire, England. This 220-hectare reserve contains heathlands, bogs, and 80 hectares of wood pasture, dominated by beech and oak pollards; many of the individuals are thought to be 400 years old or more. In the mid 19th century pollarding ceased for a period of 160 years, and grazing animals were removed in the mid 20th century. Management in the Burnham Beeches has focused on preservation of veteran trees, as most of the older beeches are hollow from decay. The City's stated goal is to structurally improve 50 trees per year. This involves the use of a professional arborist who can be responsible for reducing the size and instability of the trees and is a considerable investment. In addition to tree maintenance, 500 new pollards have been created in the reserve. Prior to embarking on this program, reserve managers conducted experimental treatments to younger, uncut beech trees in order to discern the best conditions for securing new growth from beech trees on this site. By examining the light environment, balance of branches, and cutting method, they have developed techniques that can stimulate growth in specimens over a century old. Managers have also reintroduced light grazing to open the shrubby understory with traditional methods, and they have instituted a monitoring program for vegetation and ground invertebrate populations. Because visitors from nearby London frequent the Burnham Beeches, safety is also a factor in their preservation efforts. Pruning has been conducted on hazard trees when necessary. Managers of the Burnham Beeches involve and benefit from the commitment of interest groups and volunteers, including their own volunteer program for site maintenance, private donors, the Woodland Trust, and others. The Reserve has twelve full time staff and a comprehensive management plan (Read and Frater, 1999; Barnard et al, 2005, Read et al, 2007).

The methods at Burnham Beeches focus on individual tree preservation because the resource is limited in extent and particularly fragile. Furthermore, managers have emphasized not only the product - an extremely aged and gnarled tree - but also the process: the continuous cultivation of the pollards. Their self-described purpose is "to manage the veteran trees, semi-natural ancient woodland, heath and mires in order to perpetuate them and their associated features for the future [and to] encourage public access and involvement and foster greater understanding of the site" (Read and Frater, 1999). These management methods would be suitable on a small scale in locations where preservation of individual remarkable trees or sites is desired. Sites accessible by road and those frequented for their impressive stature would also be candidates for similar approaches: within Aiako Harria, both Castillo del Ingles and Oieleku may be suitable. However, these methods are very labor intensive and demand great care and time in their execution. The cooperation of an interested public has been a key ingredient for Burnham Beeches and is a necessary ingredient even in a complex plan of "tree surgery" such as the one underway at the Reserve (Barnard et al, 2005; Read, 2007). Management of exceptional trees through careful arborist work and public understanding of the importance of the trees has been successful, and Park managers in the Basque Country can exchange techniques and information with their colleagues in pollard preservation.

Woodland and Landscape History in the United Kingdom

The Burnham Beeches is only one example of many woodland sites in Britain protected for coincident cultural and ecological values. The preservation of historically-managed landscapes is a well-established institution in the United Kingdom, as exemplified not only by pollarded trees, but also by copses, wood-pastures, hedgerows, meadows, and entire pastoral landscapes (Rackham, 1990). This work is generally incorporated within the practice of countryside management, and itself has a long history of integration with conservation efforts. In the 1860s and following, The

Commons and Footpaths Preservation Society promoted conservation of the New and Epping Forests for their outstanding history, resulting in the preservation of "Ancient and Ornamental" woodlands that had been managed in coppice systems for centuries (Tubbs, 1964; Peterken, 1996). In fact, woodlands' cultural value has been a dominant force in the establishment of reserves throughout the U.K., and continues to inspire local involvement in their study and maintenance (Rackham 1990). Additionally, the benefits of perpetual human involvement with forest ecosystems have been extensively documented by ecologists with an interest in landscape heterogeneity and the protection of rare habitats in these ancient inhabited landscapes (Peterken 1977, Read and Frater 1999). For the last three decades, managers of ancient woodlands have strived to incorporate both environmental and historical preservation goals into their activities. Their ample documentation of these endeavors can serve as a resource for park managers in similar undertakings.

Culturally Modified Trees

Culturally modified trees, (CMTs), are trees that have been modified by traditional cultural practices that frequently result in particular distinguishing features (Ostlünd 2002). The most frequently cited practices resulting in the creation of CMTs across a number of regions and societies include bark stripping for fibers, building materials, and use of the cambium layer as a food source. Scarring and burning of tree trunks was practiced by some indigenous groups to mark boundaries, while trunk carvings and notches indicate placement of natural shelves laid between trees to dry hay and fodder (Mobley 1992; Zackrisson 2000; Mallea-Olaetxe 2000). Pollarding can be viewed as another form of cultural modification for utilitarian purposes. CMTs are useful today in helping us learn about the diets, materials, migrations and sacred rituals that defined ancient cultures, including those that are no longer in practice (Prince, 2001).

The Basques in particular have been linked to various CMTs, such as arborglyphs, used by shepherds to communicate both in the Spanish Basque country and in Basque farming and herding communities of the American West (Mallea-Olaetxe, 2000). The pollard beech forests of the Basque country share many characteristic features with other regions where CMTs are common: they have been forested continuously and used by native people in the region, including contemporary farmers (Ostlünd, 2002). Like many other CMTs, pollards have a long history of subsistence and commercial use. Though few studies have been conducted specifically on the loss of CMTs, recent research indicates a decline of this resource in the 20th century. This coincides with declines in the modified trees' economic importance, and much of their disappearance is attributable to harvesting activities unrelated to earlier cultural practices. In British Columbia, trees registered as CMTs are protected by complex laws. Trees may not be logged if they are dated older than 1846 and registered as CMTs, and some First Nation management plans in this area claim a 20-30 meter protective zone around the tree (Perreault, 2002). The pressures on CMTs include both logging and management that encourages removal of aged or "damaged" trees. Scandinavian forest management, for example, emphasizes this practice, although some proponents attest that the presence of CMTs within a forest reserve indicates that it should be regarded and managed as a cultural landscape (Ostlünd, 2002). In Sweden there was little awareness or prioritization of CMTs by authorities responsible for forest management and attention focused instead on the actual archaeological remains rather than the culturally modified ecosystem and trees themselves (Ostlünd, 2002). However, increased interest in the history of Scandinavian forests has resulted in more information campaigns and government sponsored programs aimed at protecting the cultural values of the forests themselves. In the Basque Country, as in Sweden, there are no specific protections or designations for CMTs within forest reserves or elsewhere. The Swedish model of an advocacy-based, public information campaign may serve as an implementable model to redefine public perception of forests as a shared cultural as well as natural heritage.

UNESCO World Heritage Sites, World Heritage Forest Programme and Biosphere Reserves

UNESCO is one of the premiere organizations promoting the management and protection of natural and cultural heritage. Since 1972, when UNESCO adopted the Convention Concerning the Protection of the World Cultural and Natural Heritage, World Heritage sites have promoted international cooperation in the conservation of the world's cultural and natural heritage by encouraging participation of local populations in the preservation of designated sites. Today, the World Heritage list includes 936 properties, of which 725 are classified as cultural, 183 natural, and 28 mixed, amongst 153 of the 187 States Parties that have ratified the World Heritage Convention. Two of the 28 mixed natural-cultural World Heritage Sites are located in Spain, Ibiza and Pyrénées at Monte Perdido. The latter, established in 1997, is a transboundary site shared between Spain and France, and parallels Aiako Harria in its representation of threatened traditional forest and agricultural practices (UNESCO WHC). UNESCO also leads the World Heritage Forest Programme, a system of 104 forest sites that together account for some 76 million ha and 13% of all IUCN protected forests (DuVal and Crane, 2011). The World Heritage Forest Programme includes the Primeval Beech Forests of the Carpathians in Slovakia and Ukraine, granted universal importance as a genetic reservoir of beech species and other species dependent upon these forest habitats (UNESCO WHFP). Biosphere Reserves, maintained through UNESCO's Man and the Biosphere Programme, has the proposed mission of promoting an interdisciplinary research agenda and building capacity to improve the relationship of people with their environment around the world (UNESCO MAB). Since the early 1970s the Man and the Biosphere Programme has targeted the ecological, social, and economic dimensions of biodiversity loss by using a network of Biosphere Reserves as a platform for knowledge sharing, research and monitoring, education and participatory decision-making. Within this program, an area of 220 km^2 situated on the Bay of Biscay was designated the Urdaibai Biosphere Reserve in 1984 in recognition of the unique Cantabrian holm oak forests (*Quercus ilex* L.), an abundance of vascular plants and birds, and various historical artifacts from the prehistoric, Roman and Middle ages. It is currently implementing a management, land use and development plan promoting tourism, agriculture and forestry (UNESCO MAB2).

We highlight these UNESCO programs as important early contributions to a socio-ecological approach in managing and protecting resources, programs that have recognized the world heritage value of forest systems similar in nature to the pollarded beech and oak forests of Aiako Harria. However, we must emphasize that these examples are offered as a model and a framework but that management solutions must incorporate good governance at all scales. Designation from international agencies like UNESCO may bear cultural and political sensitivities if the autonomy of local authorities is bypassed in the process. While Urdaibai is the only Biosphere Reserve in Basque country, it is a rare example of a Biosphere Reserve that is not also a government-designated protected area. Local biologists point out that plantations of exotic species like eucalyptus and Monterey pine are extensive within the "core" area. While collaboration at multiple levels of government can facilitate a consistent support network for conservation efforts, local or municipal programs should have the prerogative to advance and contribute feedback to designations and management within their purview.

Finca De Fitor

The Fitor Estate, located in the Area of Natural Interest (EIN) Gavarres, Catalonia, is a Natura 2000 site recognized as an example of good management practice (Natura 2000 Networking Programme, 2005). Fitor is a 1300 hectare private holding managed by the Botey Serra family for multiple uses. The location is home to a variety of Species of Community Interest, including some of the same invertebrates found in Aiako Harria Natural Park. A featured species in their management plan is the two-tailed pasha, (*Charaxes jasius* L.), which is the largest butterfly in Europe and feeds on strawberry madrone (*Arbutus unedo* L.) trees. The estate has more than 650 species of plants and is situated at an ecological crossroads: the intersection of Mediterranean, Atlantic, and Euro-Siberian biogeographic regions. Laurels (*Laurus nobilis* L.), introduced by monks more than one thousand years ago, have naturalized along the stream corridors. The estate also has preserved and restored a number of historic structures, including the Mas Plaja house, which has been continuously occupied since the 13th century. Tile ovens and ice pits from the Middle Ages and traces of Roman lead and silver mines are distributed on the forested hills. Remnants of even earlier cultures are found in numerous grave markers and upright stones of prehistoric origin.

Beginning about 700 years ago, locals began to harvest bark from the cork oaks (*Quercus suber* L.), and have managed the landscape for cork production ever since. Other commercial enterprises utilize the natural forests and fertile soils through commercial cultivation of trees and ornamental greenery, primarily strawberry madrone and eucalyptus trees. The estate also earns some revenue from use as a tourist retreat and private hunting ground. Within the past 100 years, depression, civil war, and a forest fire have ravaged the productivity of the forest. When the estate was acquired around 1950, many of the trees were 200 or more years old and failing in their production of cork. Since that time the family has used the estate as extensive experimental grounds, inviting scientific study and carrying out their own commercial and forestry trials. Meanwhile, they have rehabilitated the oak forests to now produce 100 tons of cork annually, about 2% of the annual Catalan supply. On the whole, the family sees the estate's purpose as "enriching the cultural and natural landscape of Europe" (Botey Serra 2004). Management strategies are geared toward multifunctionality: cork production, live nursery products, historic preservation, and tourism form an interrelated network of revenues and services. The Fitor Estate features ancient monuments, modern forestry, and regionally important biodiversity. While private ownership facilitates a highly flexible and adaptable standard of management that is harder to achieve in a protected area of more complex governance, the model of traditional practice combined with interpretation and commercial activity is worth consideration.

RECOMMENDATIONS

Continued Inventory and Study

The first recommended action is the development of a baseline inventory of the pollarded forests, as preservation experts have already suggested (Read, 2004). Current understanding of the extent of these beech pollard forests and their relative importance is based on an estimate of the area that they cover, which has been cited as 800 ha (Read, 2006). However, there has not been any comprehensive pollard inventory conducted to date (I. Aizpuru, personal communication, October 19 2009). An accurate, up-to-date characterization of the pollarded trees in the park, including number, exact location, density, and condition, would improve ongoing management of the resource in a number of ways. First, it would establish the quantity and record the location of these forests and trees, providing a ready reference for public information and strengthening the case for additional and ongoing funding by confirming their global uniqueness. Secondly, a survey of pollards would help categorize stands of forests based on their condition as 1) those which are at high risk of collapse; 2) those which would yield the most benefits from conversion to maiden beech stands; and 3) those which might be managed using traditional methods. This would help park managers prioritize areas for treatment, experimentation, or preservation. Managers should engage local stakeholders including park users and local schools or universities in the inventory process. For example, members of the Mountaineering Federation regularly hike the region and take great pride in the park's natural resources and beauty; engaging park usergroups can help connect participants relate more directly to the cultural as well as natural components of the landscape, while minimizing the cost of operations and maintenance for the park. A participatory approach to inventory has been successfully utilized by urban forestry groups and serves the dual function of teaching critical science and job skills as well as promoting proficiency with GPS units and other survey and monitoring equipment (Margerum, 1999; Alexander, 2008).

A complete inventory of the pollards could help identify trees that may be suitable for landmark status as a Singular Tree. Singular Trees are designated by the Basque government based upon their extraordinary or noteworthy characteristics such as size, age, history, beauty, or situation, and are afforded special protection. Currently there are 26 such trees in the country, 11 of which are in Gipuzkoa (Basque Government, 2006). Designating a few selected pollard beeches as Singular Trees could generate visibility and awareness of the unique history that gave rise to the pollards, which would benefit all pollarded beeches and oaks within the park.

Finally, it will be useful to continue investigation of ecological processes in the pollarded forests. Park managers have already begun this process by conducting trials in establishing regeneration where senescent pollards have provided an open understory, and through extensive study in forest diversity (Carte 2008, Etayo 2008, I. Aizpuru, personal communication, October 19 2009). Next steps might include a study of gap dynamics and composition of stands with focus on the regenerating forests that will gradually replace the senescing pollards, and characterization of the pollard forest as a human-created plant community for management and education purposes (Whitney and Adams, 1980; Marcucci, 2000; Rozas, 2004). Further study and data collection will help to inform the long-term plans for management of the forest.

Public Awareness and Education Campaign

Complementary to a thorough inventory, a targeted community-outreach campaign is needed to highlight the local cultural heritage and global importance of the Basque pollarded forests. Our own limited interactions with park user groups, such as a group of mountaineers from the nearby city of Bilbao, supported statements made by park staff that there was little public awareness as to how these trees had been formed and their specific relevance to Basque history in the region. Indeed, many of the local park users had grown up familiar with the common spectacle of the pollards and consequently gave them little thought. However, in recent years, park rangers have noted that foreign tourists in particularly are expressing specific interest in the pollards.

This is likely a consequence of the widespread publicity efforts amongst ancient tree enthusiasts through forums such as the Ancient Tree Forum, particularly in Great Britain. Signage of any sort relating to the pollards was virtually absent from the park at the time of our assessment. The only visual media linking the present form of the pollards to their traditional uses in charcoal making seen during our trip was an informational panel in the museum of the Basque Center for Biodiversity located in Busturia, 125 km from Aiako Harria, although we were informed that closer panels exist. Increasing signage in high traffic and heavily pollarded regions of the park such as the picnic area at Oieleku, around the Castillo del Ingles, and at trailheads such as the Arritxulo hostel area, would provide a cost-effective and efficient means of educating park users about the importance of pollards in Basque history.

Interactive modes of outreach and education should be coupled with passive methods of public information like signage. We recommend that teams of arborists and the remaining traditional charcoal makers collaborate by organizing public demonstrations in the park to ensure that this unique form of local knowledge is not lost to future generations. Demonstration charcoal mounds have already been suggested as another potential activity with both local and tourist appeal (A. Elosegi, personal communication, October 21 2009).

Multiple Use Management

Basques have a long tradition of harnessing the economic potential of working forests. This has left an indelible mark upon the landscape, exemplified most recently by the plantations of Monterrey pine (*Pinus radiata* D.Don.), red oak (*Quercus rubra* L.) and other exotic species that dominate the park and surrounding lands today. Accordingly, the pollard forests should be utilized in a number of ways to promote biodiversity, cultural awareness, scientific progress in the management of veteran and lapsed trees, and even generate revenue where possible.

We therefore recommend that following the inventory, various pollarded areas be ascribed different uses and purposes. For example, continued management and pollarding would not be practical on some of the steepest slopes where they occur, but the dead wood from pollard boughs could provide an important source of slope stabilization and erosion control while encouraging understory and herbaceous growth by protecting regeneration from grazing pressure. Some of the best-preserved sites, particularly those surrounded by other cultural relics like the cromlechs near Oieleku and Castillo del Ingles, should be left as "museum sites" for public enjoyment and recreational use. These sites should be highly promoted to tourists and hikers, and would make great locations for multi-language informational signs.

Other less accessible sites should be selected as experimental plots to learn more about management and pruning responses, and the environmental heterogeneity of the terrain should be utilized to provide lessons for the management of veteran trees under a great deal of different conditions including slope, aspect, and tree age.

Finally, some areas should be preserved for demonstrations of traditional pollarding and charcoal making practices.

Integrative Cultural-Biological Framework for Valuing Pollards

A largely biodiversity-centric justification for pollard conservation was employed by park managers in accordance with the LIFE project grant, which lasted from 2006 to October of 2009. Natura 2000 does not differentiate between the value of natural landscapes versus those created or modified by human influence, although the former dominate its preservation efforts. Although the classification of the pollards as a Natura 2000 site has granted them increased visibility and targeted them for continued research, the result has been that the pollards are valued largely for their contributions to the region's invertebrate biodiversity. If this management value were the only one in consideration, it could be reasoned then that the best management approach would simply be to leave the trees to lapse and break, creating a broader mosaic of dead and dying wood types to promote habitat formation and biodiversity. This is a hypothetical extreme, but nonetheless illustrates the dangers in assuming too narrow a framework for justifying a course of management actions. In a discussion of preserving the "semi-natural" or human-modified landscape for its natural values, Alexander (2008) explains this dichotomy.

If we were to attempt to apply the accepted meaning of "wilderness" to the Old World, it can readily be seen that no place can claim total isolation from human influence.

THE LIFE + POLLARD PROJECT

Although the 2010-2013 Life + project is still in progress, the details and actions to date indicate a project and approach closely aligned with the values and recommendations of this paper. The justifications for a Life project exclusively treating beech pollards, (Table 1) are split between values that touch upon the ecological importance, cultural heritage, and global rarity of the pollards, as well as the importance of immediate action for conservation.

Similarly, the actions that have been proposed range widely in nature from biodiversity monitoring goals reminiscent of the prior Life+ Nature project to ambitious unprecedented efforts to document and transmit traditional knowledge (Table 2).

This classification system represents the author's analysis of the frameworks by which the pollard beeches have been valued and classified as part of the Life+ Project. After assigning values, the Actions and Justifications broke down as follows: Conservation (13); Ecology (11); Cultural Heritage (8); Outreach and Education (6); Global Rarity (4).The 2010 Life+ Pollard Project assumes a diverse and integrative approach to addressing the multiple management concerns of the beech pollards in Aiako Harria Natural Park. The Actions and Justifications of this project that were disclosed by the time of this writing are aligned with key components articulated in this paper; for instance, continued inventory, public awareness and environmental educational, and some multiple use management. Finally, an integrative cultural-biological framework for valuing pollards is exemplified by the diversity of values identified. The actions and justifications listed here represent an initial synopsis and are not necessarily comprehensive of the project aims, leaving room for further integration of some of the recommendations made in this article may have been incorporated since the 2010-2013 Life + Project description.

While many action points are included specifically for public outreach, there is still no mention of improving awareness and visibility of pollards and their unique history within the park itself, where park user groups would be most efficiently accessed. We strongly recommend therefore that signage be increased within the park in addition to the other means of exposure. Secondly, there is an enormous quantity of pollarded individuals over a large area, estimated at 150,000 modified beeches, oaks and ashes over 800 ha. (I. Portu, personal communication January 15th 2010).

Table 1. Justifications of a Life Project on Beech Pollards

	Justification	Values
1	Habitat of vital importance for the conservation of biodiversity/ ecological role of individuals	Ecology
2	Habitat threatened with extinction in the short-medium term.	Ecology/ Conservation
3	Global Rarity- major concentrations of old beech pollards found only in Romania, England, Hungary and northern Spain	Global Rarity
4	Important part of the heritage and cultural legacy of the Basque Country, and Europe	Cultural Heritage/ Global Rarity
5	Limited timescale for action	Conservation
6	Better understanding of distribution and ecology of habitat needed to inform management decisions	Ecology/ Conservation
7	Important roles as connection habitats	Ecology
8	Importance of raising public awareness as to the historical, natural and cultural heritage represented by trees	Cultural Heritage/ Outreach and Education
9	Informing private land owners of the uniqueness	Global Rarity/ Cultural Heritage/ Outreach and Education
10	Preserving a native Western Europe heritage	Cultural Heritage
11	Only the Basque country still harbors living individuals with the traditional knowledge of management- valuable cultural legacy	Global Rarity/ Cultural Heritage
12	LIFE+ the most powerful instrument in carrying on the tradition of knowledge, management	Conservation
13	Lack of other conservation initiatives in this area	Conservation
14	Geographic importance of Basque Country as a connecting territory	Ecology (Landscape)

Source: I. Portu, personal communication, January 15th 2010).

According to the actions, only 162 ha have been ascribed a specific ecological or conservation purpose, leaving many more hectares of pollarded stands available for use in recreation or experimental treatments for managing senescing pollards. To have such a surplus of candidate trees available is a great resource and can progress scientific knowledge on the management of structurally unstable pollards while involving the public through engagement of local schools and universities. Finally, despite the wide range of social and biological values cited in our analysis of the project's actions and justifications, very few points actually combined values of social and biological nature for the purpose of management. The action that best embodied an integrative approach was the documentation and transfer of tree pollarding techniques to help sustain habitat features. In many ways, this action summarizes what we see to be the enormous potential of this project to link social and ecological values and research in a meaningful and transferable way. This point makes the direct link between traditional human intervention and habitat stability, both structural and compositional, as being not only mutually compatible, but worthy of further study, understanding and perpetuation through the transfer of such techniques to future generations.

Table 2. Proposed Actions of the 2010 Life+Pollard Project

	Action	Values
1	Protection of 12 ha of pollard forests to guarantee saproxylic habitat	Ecology
2	Improvement of habitat features on 150 ha of acidophilic beechwood in Gipuzkoa	Ecology
3	Increased knowledge on biodiversity values, management requirements of pollarded trees	Ecology
4	Public acquisition of areas hosting target species	Conservation
5	GIS mapping of priority habitats in project area	Conservation
6	Documenting and transfer of tree pollarding techniques to help sustain habitat features	Cultural Heritage/ Ecology/ Conservation
7	Elaboration and implementation of a long-term management strategy for pollarded trees in the project area	Conservation
8	Creation of a European network of forestry habitats with pollarded trees supporting saproxylic spp	Ecology/ Conservation
9	Cataloguing invertebrate species, preparing management plans	Ecology/ Conservation
10	Creation of a web, informative bulletins, press conferences, news in specialized publications	Outreach and Education
11	Talks, trainings and regional and international technical conferences	Outreach and Education
12	Digital cartography of pollarded copses, interest at appropriate management scale	Conservation
13	Training new generations of pollarders	Cultural Heritage/ Outreach and Education
14	Collection of traditional and modern pollarding techniques in a Good Practice Guide	Cultural Heritage/ Conservation/ Outreach and Education

Source: I. Portu, personal communication, January 15th 2010.

CONCLUSION

Efforts to protect the cultural landscape often overlook the natural components that make them unique. This is a loss for cultural preservation as well; as Aldo Leopold explains, "The rich diversity of the world's cultures reflects a corresponding diversity in the wilds that gave them birth" (1949) (quoted in Alexander, 2008). The Basque Country is a region rich in cultural and natural resources, which are epitomized by the remaining pollard beech forests of Aiako Harria. The year 2010 was declared by the UN to be the International Year of Biodiversity, and the year 2011 has followed as the UN International Year of Forests with the theme "Forests for People." The pollarded forests of Aiako Harria, along with the various case studies referenced in this paper, remind us of the long legacy forests have in serving the eclectic needs and values of people, and the reciprocal obligation that people hold to manage them wisely and sustainably. Though enjoyed and admired locally for many years, the ecological importance and global uniqueness of the pollarded trees of Aiako Harria are only beginning to come into focus. The forests they characterize are both for and by people; through versatile management, the same individual trees have remained relevant to the needs and values of changing societies for over two hundred years. Never before has public interest in the connections between sustainable resources use, biodiversity and culture been higher. Aiako Harria provides an ideal opportunity for furthering discussions and promoting management practices favorable to preserving biocultural diversity.

ACKNOWLEDGMENTS

The authors would like to acknowledge Tamara Muruetagoiena for her generous assistance and guidance throughout the project.

REFERENCES

Alexander, M. (2008). Management Planning for Nature Conservation: A Theoretical Basis and Practical Guide. The Netherlands: Springer Science +Business Media B.V.

Altuna, J. and Areso, P. (1977). Excavaciones en los cromlechs de Oyanleku (Oyarzun, Guipúzcoa). *Munibe*, 29, 75-76.

Aragon Ruano, A. (2001). El bosque guipuzcoanoen en la Edad Moderna: Aprovechamiento, ordenamiento legal y conflictividad. *Munibe,* Suplemento 14, 285pp. San Sebastián: Sociedad de Ciencias Aranzadi.Barandiarán, I. (1973). Irún romano. *Munibe*, 25, 19-28.

Basque Government 2006. Basque Country Network of Protected Natural Spaces: Aiako Harria.

Benito, A. M. (1988). Cerámicas del yacimiento submarino del cabo de Higer (Hondarribia), *Munibe*, 40, 123-163.

Barnard, A., Hartup, M. and Read, H. (2005). *Burnham Beeches Management Plan.* April 2005-March 2010 Full plan. City of London. Retrieved October 1st, 2009 from http://217.154.230.218/NR/rdonlyres/3560F960-F4D4-4F35-AE82-57B80D72816C/0/OS_BB_manageplan0510.pdf.

Botey Serra, J. (2004). Finca de Fitor. The Conservation of Europe's Natural and Cultural Heritage through Estate Mangement by Family Forestry. Retrieved October 1st, 2009 from http://www.natura.org/natura2000management/doc/es_fitor_sitedescription.pdf.

Butler, J. (2006). Ancient Pollards and Maidens - securing their future. Paper presented at the First European Colloquium on Pollards, Vendôme, October 26-28 2006. Retrieved October 10 2009 from http://www.maisonbotanique.com/dyn/17acte_8_butler.pdf.

Carte, S. P. (2008). Inventario y seguimiento de la entomofauna del hayedo de Oieleku (Oiartzun, Parque Natural de Aiako Harria). *LIFE/Natura 2000 report.* Retrieved October 1st, 2009 from http://www.lifeaiakoharria.net/datos/documentos/insectos%20Oieleku2007.PDF.

Christensen, M., Heilmann-Claussen, J., Walleyn, R. and Adamcik, S. (2004). Wood Inhabiting Fungi as Indicators of Nature Value in European Beech Forests. In M. Marchetti (Ed), *Monitoring and Indicators of Forest Biodiversity in Europe–From Ideas to Operationality.* (pp. 229-238). European Forest Institute Proceedings No. 51.

Department of Environment and Regional Planning. Retrieved October 1st, 2009 from http://www.ingurumena.ejgv.euskadi.net/r49-4975/en/contenidos/informacion/enp/en_1086/aia_normativa_i.html.

DuVal, A., Crane, P.R. (2011). World Heritage sites and the future of forests. World Heritage Review 61, 8-17.

Etayo, J. (2008). Estudio de los líquenes y hongos liquenícolas de Aiako-Harria. *LIFE/Natura 2000 report.* Retrieved October 1st, 2009 from http://www.lifeaiakoharria.net/datos/documentos/Estudio%20de%20los%201%C3%ADquenes%20de%20AH_2008.pdf.

European Commission. (2008). Commission Decision of 12 December 12 2008 adopting, in accordance with Directive 92/43/EEC, a second list of updated sites of Community importance for the Atlantic biogeographical region. *Official Journal of the European Union,* C (2008) 8119.

Fenley, J. (1950). Pollarding: Age-Old Practice Permits Grazing in Pays Basque Forests. *Journal of Range Management*, 3, 316-318.

Ferrini, F. (2006). Pollarding and its effects on tree physiology: a look to mature and senescent tree management in Italy. Paper presented at the First European Colloquium on Pollards, Vendôme, October 26-28 2006. Retrieved October 10 2009 from http://www.maisonbotanique.com/dyn/17acte_10_ferrini.pdf.

Franklin, J., Shugart, H. and Harmon, M. (1987). Tree Death as an Ecological Process. *BioScience*, 37, 550-555.

Garcia-Ruiz, J. M., Lasanta, T., Ruiz-Flano, P., Ortigosa, L., White, S., Gonzalez, C., and Marti, C. (1996). Land-use changes and sustainable development in mountain areas: a case study in the Spanish Pyrenees. *Landscape Ecology*, 11, 267-277.

General Directorate for Woodlands and the Natural Environment. (2005). *LIFE-Nature Project 2005-2009: Conservation and Restoration of Aiako Harria SCI (LIFE05 NAT/E/000067).* Retrieved October 1st from http://www.lifeaiakoharria.net/datos/documentos/forest_portu.pdf.

Grove, S. (2002). Saproxylic Insect Ecology and the Sustainable Management of Forests. *Annual Review of Ecology and Systematics*, 33, 1-23.

Kenk, G. and Guehne, S. (2001). Management of transformation in central Europe. *Forest Ecology and Management,* 151, 107-119.

Kurlansky, M. (1999). The Basque History of the World: The Story of a Nation. New York: Penguin Books.

Loidi, J. (2005). The Cantabrian-Atlantic oak and beech forests: Human influence throughout history. *Botanika Chronika*, 18, 161-173.

Mallea-Olaetxe, J. (2000). *Speaking through the aspens: Basque tree carvings in California and Nevada.* University of Nevada Press, Reno and Las Vegas. 237 pp.

Marcucci, D. J. (2000). Landscape history as a planning tool. *Landscape and Urban Planning,* 49, 67-81.

Margerum, R. (1999). Integrated environmental management: the foundations for successful practice. *Environmental Management*, 24, 151-166.

Ódor, P. and Standovár, T. (2001). Richness of Bryophyte Vegetation in Near-Natural and Managed Beech Stands: The Effects of Management-Induced Differences in Dead Wood. *Ecological Bulletins,* 49, 219-229.

Onaindia M. and Mitxelena, A. (2009). Potential use of pine plantations to restore native forests in a highly fragmented river basin. *Annals of Forest Science,* 66, 305.

Peterken, G. F. (1977). General Management Principles for Nature Conservation in British Woodlands. *Forestry* Vol. 50 No. 1 197.

Peterken, G. F. (1996). Natural Woodland: Ecology and Conservation in Northern Temperate Regions. Cambridge University Press, Great Britain.

Petit, S. and Watkins, C. (2003). Pollarding Trees: Changing Attitudes to a Traditional Land Management Practice in Britain 1600–1900. *Rural History*, 14, 157-176.

Perreault, P. A. (2002). Culturally Modified Tree (CMT) Management in Niaka´pamux Territory: Shaping First Nations Participation through Consultation. Master of Science Thesis, The University of British Columbia, Vancouver, British Columbia.

Rackham, O. (1990). *Trees and Woodland in the British landscape*. Revised edition. J. M. Dent and Sons Ltd., London.

Read, H. J. and Frater, M. (1999). *Woodland Habitats*. Routledge, New York.

Read, H. J. (2004). A study of practical pollarding techniques in northern Europe: Report of a three month study tour, August to November 2003. Retrieved October 1st, 2009 from http://frontpage.woodland-trust.org.uk/ancient-tree-forum/atfinternational/images/1%20Front%20Acknowledgements%20and%20contents.pdf.

Read, H. (2006, October). A brief review of pollards and pollarding in Europe. Paper presented at the First European Colloquium on Pollards, Vendôme, October 26-28 2006. Retrieved October 1st 2009 from http://www.maisonbotanique.com/dyn/12acte_2_read.pdf.

Read, H. (2007). Pollarding Beech trees in the Basque Country: report of work carried out in February 2007. City of London, Leitza Town Council (Navarra) and Aiako Harria Natural Park (Gipuzkoa).

Read, H., Forbes, V., Young, J. (2007). Specialist survey of all old pollards at Burnham Beeches and Work Programme for 2007/8 to 2015/06. Retrieved October 1st, 2009 from http://www.pro-natura.net/Pollards-Burnham-Beeches.pdf.

Rescia, A. J., Schmitz, M. F., Martin de Agar, P., De Pablo, C. L., and Pineda, F. D. (1997). A Fragmented Landscape in Northern Spain Analyzed at

Different Spatial Scales: Implications for Management. *Journal of Vegetation Science*, 8, 343-352.

Rozas, V. (2003). Regeneration patterns, dendroecology, and forest-use history in an old-growth beech–oak lowland forest in Northern Spain. *Forest Ecology and Management* 182, 175-194.

Rozas, V. (2004). A dendroecological reconstruction of age structure and past management in an old-growth pollarded parkland in northern Spain. *Forest Ecology and Management* 195, 205-219.

Schmitz, M. F., Atauri, J. A., De Pablo, C. L., De Agar, P. M., Rescia, A. J., and Pineda, F. D. (1998). Changes in land use in Northern Spain: Effects of forestry management on soil conservation. *Forest Ecology and Management* 109, 137-150.

Segura, F. (2009). Hemos pedido ayuda a la UE para salvar las hayas trasmochas. Diario Vasco May 27, 2009.

Tubbs, C. R. (1964). Early Encoppicements in the New Forest. *Forestry*. 1964 37(1):95-105.

UNESCO Man and the Biosphere Programme. Accessed December 2009 from http://www.unesco.org/new/en/natural-sciences/environment/ecological-sciences/man-and-biosphere-programme/.

UNESCO World Heritage Convention. Accessed December 2009 from http://whc.unesco.org/en/list.

UNESCO World Heritage Forest Programme. Accessed December 2009 from http://whc.unesco.org/en/forests.

Whitney, G. and Adams, S. D. (1980). Man as a maker of new plant communities. *Journal of Applied Ecology* 17, 431-448.

Zackrisson, O., Östlund, L., Korhonen, O. and Bergman, I. (2000). The ancient use of *Pinus sylvestris* L. (Scots pine) inner bark by Sami people in northern Sweden, related to cultural and ecological factors. *Vegetation History and Archaeobotany* 9, 99-109.

In: Multiple-Use Management ... ISBN 978-1-61942-328-2
Eds: A. DuVal, T. Muruetagoiena et al.

SYNTHESIS

In a public park like Aiako Harria, management goals and objectives are necessarily multi-tiered. Conservation and restoration efforts must be compatible with stakeholder needs in order to gain public support. Projects that affect and alter the landscape in public parks are particularly challenging; emerging paradigms in ecological restoration are often at odds with cultural perceptions of land use and aesthetics. However there are opportunities at all stages of the process to engage the public and build supportive and sustainable relationships between the public and the park.

Several re-occurring themes in the chapters of this compilation highlight current directions in resource management. Most of the chapters address the need for a transition from mono-dominant land uses to more holistic management models. In his analysis of watershed restoration, Blazewicz (Chapter 2) recommends building a "green corridor" with broad connectivity between riparian zones. In doing so, the park could meet goals of promoting biodiversity, ecological restoration, and recreation by creating corridors for wildlife, limiting erosion and developing routes for hikers and bikers with continuity from headwaters to sea level. Cousins and DuVal (Chapter 5) recommend targeting different stands of the senescing pollards for specific objectives appropriate to the site and current needs by delineating certain areas of pollarded forest for tourism, research and education. Adams and Lawhon (Chapter 1) outline steps needed to achieve the goal of integrated, multiple-use management; they recognize the importance of diversifying management objectives but argue that successful operationalization will require better outreach and engagement of the parks public.

Engagement and integration of the public into park processes is a constant thread of discussion in the papers. The necessity of a public-driven planning process is present in all recommendations, regardless of the management objective. Caligiuri (Chapter 3) reminds us that at the root of any forest management process lies "the social values and objectives of what is being managed for and why"; public participation at all levels of decision-making ensures that actions are carried out in accordance with stakeholder values. Adams and Lawhon (Chapter 1) outline several frameworks and toolsets, such as systematic problem orientation, social process mapping and the Social Multi-Criteria Evaluation method, to identify values and conflicts that affect the decision-making process.

Engagement of the public goes beyond building consensus; there should be more direct involvement of the public in park processes as well. Adams and Lawhon (Chapter 1) note the decline of community volunteerism in the park; local residents helped to clear trails and control burns in grazing areas on a regular basis when agrarian livelihoods predominated around the park. There is room for further efforts in outreach and communication to transfer an understanding and appreciation of restoration efforts to a public rooted in a history of highly modified agrarian landscapes. The park struggles with a public that still sees dead wood and dense undergrowth, important for structural complexity, biodiversity and forest succession, as signs of a "dirty" and neglected woodland. The importance of communication and transference of knowledge and practice goes both ways. Cousins and DuVal (Chapter 5) illustrate a waning tradition of local resource management that could be lost without proper documentation and transfer of the knowledge. Adams and Lawhon (Chapter 1) suggest the revival of community work days to bring people together over park projects. They also recommend that the park sponsor educational forums to inform the public about park management issues such as biodiversity conservation, erosion control, and invasive species. While public discussions are held, such as the one we participated in on forest management for water quality, improving attendance is important.

The reports also highlight the importance of working with rather than against natural processes. This means developing solutions appropriate to the site and scale of management and planning along a longer term than the time frames dictated by political terms and funding cycles. Caligiuri (Chapter 3) reviews the region's history of place-based agriculture utilizing a heterogenous mosaic of microsites, and recommends that foresters adopt the same sensibilities by concentrating timber production onto only the most productive and accessible sites with the highest potential returns. Cousins and DuVal (Chapter 5) recommend that steep and eroding hillsides be allowed to revert to maiden beech forests, with trailside or heavy traffic areas of the park targeted for continued pollarding, demonstrations and experimental plots. Building upon pre-existing natural features and processes and enhancing ecosystem services is cost-effective over the long term. The statement from the Army Corps of Engineers in 1972 that "Nature has already provided the least-cost solution to future flooding" cited by Blazewicz (Chapter 2) could be said for other management objectives as well. Caligiuri (Chapter 3) demonstrates that nature has provided the least-cost solution to restoration of native forests too, as natural regeneration abounds in the understories of exotic conifer plantations. He calls for a "close-to-nature" silviculture approach to utilize the advantages of natural processes, such as regeneration and disturbance, in balancing the restoration of native forests with the harvest of timber stands. Blazewicz (Chapter 2) makes a case for rerouting transportation and development along the contours of a watershed to minimize flooding and erosion. Lowery (Chapter 4) illustrates the unquantified benefits of ecosystem services such as biodiversity, water quality, carbon sequestration and resilience building that should be considered alongside short-term financial considerations in forest management. Lowery also urges foresters to generate more data on local market conditions, and to consider factors specific to the region such as the availability of post-harvest processing.

Increasingly, data generated by long-term studies and global initiatives such as the Millennium Ecosystem Assessment have built upon our understanding of the complexity, and necessity of ecosystem processes. Policy makers at multiple scales of government recognize the importance of environmental planning for development and long-term sustainability, and cross-sectoral partnerships allow conservation successes to become political and economic successes too. The LIFE projects within the park represent a laudable commitment to conservation and restoration efforts that can be further strengthened by incorporating the recommendations from external reviews such as ours. The chapters in this collection urge park managers to consider sites within the context of their ecosystems; make decisions that capture a time scale relevant to natural processes; always return to the core values and objectives when considering management; and make greater efforts to include the public at every stage in the process.

INDEX

A

B

C

D

E

F

G

H

I

Q

R

S

T

U

V

W

Y